Bridge Engineering

Bridge Engineering

Ramesh Shah

RANDOM

RANDOM PUBLICATIONS
NEW DELHI (INDIA)

Bridge Engineering

ISBN 978-93-5111-433-8

Published in 2014 in India by

RANDOM PUBLICATIONS

4376-A/4B, Gali Murari Lal, Ansari Road
New Delhi-110 002
Phone : +91-11-43580356, +91-11-23289044
e-mail: randomexports@gmail.com, sales@randompublications.com, info@randompublications.com

Reprinted 2026

Type Setting by: Friends Media, Delhi-110089

Preface

A bridge is a structure built to span physical obstacles such as a body of water, valley, or road, for the purpose of providing passage over the obstacle. There are many different designs that all serve unique purposes and apply to different situations. Designs of bridges vary depending on the function of the bridge, the nature of the terrain where the bridge is constructed and anchored, the material used to make it, and the funds available to build it. All major bridges are built with the public's money. Therefore, bridge design that best serves the public interest has a threefold goal: to be as efficient, as economical, and as elegant as is safely possible. Efficiency is a scientific principle that puts a value on reducing materials while increasing performance. Economy is a social principle that puts value on reducing the costs of construction and maintenance while retaining efficiency. Finally, elegance is a symbolic or visual principle that puts value on the personal expression of the designer without compromising performance or economy. There is little disagreement over what constitutes efficiency and economy, but the definition of elegance has always been controversial. The first principle holds that the structure of a bridge is the province of the engineer and that beauty is fully achieved only by the addition of architecture. The second idea, arguing from the standpoint of pure engineering, insists that bridges making the most efficient possible use of materials are by definition beautiful. The third case holds that architecture is not needed but that engineers must think about how to make the structure beautiful. This last principle recognizes the fact that engineers have many possible choices of roughly equal efficiency and economy and can therefore express their own aesthetic ideas without adding significantly to materials or cost.

Bridges can be divided into two categories: standard overpass bridges or unique-design bridges over rivers, chasms, or estuaries.

This article describes features common to both types, but it concentrates on the unique bridges because of their greater technical, economic, and aesthetic interest.

The book also includes an overview of seismic design practices in Japan and the changes in retrofit practices precipitated by that earthquake. This present book covers the entire gamut of bridge engineering in relation to basic approaches to design of bridges, supported by numerous case studies. Bridge Engineering has been thoroughly revised to reflect the changes in technology that have occurred in the past. The book has been specifically designed to suit the requirements of design and practising engineers as well as students.

I would like to thank my team for standing beside me throughout my career and writing this book. My special thanks go to "Random Publications" who have published the book.

—Ramesh Shah

Contents

1

Introduction

A bridge is a structure built to span physical obstacles such as a body of water, valley, or road, for the purpose of providing passage over the obstacle. There are many different designs that all serve unique purposes and apply to different situations. Designs of bridges vary depending on the function of the bridge, the nature of the terrain where the bridge is constructed and anchored, the material used to make it, and the funds available to build it.

***Figure:** The Akashi Kaikyô Bridge in Japan, currently the world's longest suspension span.*

The first bridges were made by nature itself — as simple as a log fallen across a stream or stones in the river. The first bridges made by humans were probably spans of cut wooden logs or planks and eventually stones, using a simple support and crossbeam arrangement. Some early Americans used trees or bamboo poles to cross small caverns or wells to get from one place to another. A common form of lashing sticks, logs, and deciduous branches together involved the use of long reeds or other harvested fibres woven together to form a connective rope capable of binding and holding together the materials used in early bridges.

***Figure:** The Arkadiko Bridge in Greece (13th century BC), one of the oldest arch bridges in existence*

The Arkadiko Bridge is one of four Mycenaean corbel arch bridges part of a former network of roads, designed to accommodate chariots, between Tiryns to Epidauros in the Peloponnese, in Greece. Dating to the Greek Bronze Age (13th century BC), it is one of the oldest arch bridges still in existence and use. Several intact arched stone bridges from the Hellenistic era can be found in the Peloponnese in southern Greece.

The greatest bridge builders of antiquity were the ancient Romans. The Romans built arch bridges and aqueducts that could stand in conditions that would damage or destroy earlier designs. Some stand today. An example is the Alcántara Bridge, built over the river Tagus, in Spain. The Romans also used cement, which reduced the variation of strength found in natural stone. One type of cement, called pozzolana, consisted of water, lime, sand, and volcanic rock. Brick and mortar bridges were built after the Roman era, as the technology for cement was lost then later rediscovered.

The *Arthashastra* of Kautilya mentions the construction of dams and bridges. A Mauryan bridge near Girnar was surveyed by James Princep. The bridge was swept away during a flood, and later repaired by Puspagupta, the chief architect of emperor Chandragupta I. The bridge also fell under the care of the Yavana Tushaspa, and the Satrap Rudra Daman. The use of stronger bridges using plaited bamboo and iron chain was visible in India by about the 4th century. A number of bridges, both for military and commercial purposes, were constructed by the Mughal administration in India.

Although large Chinese bridges of wooden construction existed at the time of the Warring States, the oldest surviving stone bridge in China is the Zhaozhou Bridge, built from 595 to 605 AD during the Sui Dynasty. This bridge is also historically significant as it is the world's oldest open-spandrel stone segmental arch bridge. European segmental arch bridges date back to at least the Alconétar Bridge (approximately 2nd century AD), while the enormous Roman era Trajan's Bridge (105 AD) featured open-spandrel segmental arches in wooden construction.

Rope bridges, a simple type of suspension bridge, were used by the Inca civilization in the Andes mountains of South America, just prior to European colonization in the 16th century.

During the 18th century there were many innovations in the design of timber bridges by Hans Ulrich, Johannes Grubenmann, and others. The first book on bridge engineering was written by Hubert Gautier in 1716. A major breakthrough in bridge technology came with the erection of the Iron Bridge in Coalbrookdale, England in 1779. It used cast iron for the first time as arches to cross the river Severn.

With the Industrial Revolution in the 19th century, truss systems of wrought iron were developed for larger bridges, but iron did not have the tensile strength to support large loads. With the advent of steel, which has a high tensile strength, much larger bridges were built, many using the ideas of Gustave Eiffel.

In 1927 welding pioneer Stefan Bry³a designed the first welded road bridge in the world, the Maurzyce Bridge which was later built across the river S³udwia at Maurzyce near £owicz, Poland in 1929. In 1995, the American Welding Society presented the Historic Welded Structure Award for the bridge to Poland.

Types of Bridges

Bridges can be categorized in several different ways. Common categories include the type of structural elements used, by what they carry, whether they are fixed or movable, and by the materials used.

Structure Type

Bridges may be classified by how the forces of tension, compression, bending, torsion and shear are distributed through their structure. Most bridges will employ all of the principal forces to some degree, but only a few will predominate.

The separation of forces may be quite clear. In a suspension or cable-stayed span, the elements in tension are distinct in shape and placement. In other cases the forces may be distributed among a large number of members, as in a truss, or not clearly discernible to a casual observer as in a box beam.

Beam bridge Beam bridges are horizontal beams supported at each end by substructure units and can be either *simply supported* when the beams only connect across a single span, or *continuous* when the beams are connected across two or more spans. When there are multiple spans, the intermediate supports are known as piers.

The earliest beam bridges were simple logs that sat across streams and similar simple structures. In modern times, beam bridges can range from small, wooden beams to large, steel boxes. The vertical force on the bridge becomes a shear and flexural load on the beam which is transferred down its length to the substructures on either side They are typically made of steel, concrete or wood. Beam bridge spans rarely exceed 250 feet (76 m) long, as the flexural stresses increase proportional to the square of the length (and deflection increases proportional to the 4th power of the length).

However, the main span of the Rio-Niteroi Bridge, a box girder bridge, is 300 metres (980 ft). The world's longest beam bridge is Lake Pontchartrain Causeway in southern Louisiana in the United States, at 23.83 miles (38.35 km), with individual spans of 56 feet (17 m). Beam bridges are the most common bridge type in use today.

Truss Bridge

A truss bridge is a bridge whose load-bearing superstructure is composed of a truss. This truss is a structure of connected elements forming triangular units. The connected elements (typically straight) may be stressed from tension, compression, or sometimes both in response to dynamic loads. Truss bridges are one of the oldest types of modern bridges. The basic types of truss bridges shown in this article have simple designs which could be easily analyzed by nineteenth and early twentieth century engineers. A truss bridge is economical to construct owing to its efficient use of materials.

Cantilever bridgeCantilever bridges are built using cantilevers—horizontal beams supported on only one end. Most cantilever bridges use a pair of continuous spans that extend from opposite sides of the supporting piers to meet at the centre of the obstacle the bridge crosses. Cantilever bridges are constructed using much the same materials & techniques as beam bridges. The difference comes in the action of the forces through the bridge. The largest cantilever bridge is the 549-metre (1,801 ft) Quebec Bridge in Quebec, Canada.

Arch bridgeArch bridges have abutments at each end. The weight of the bridge is thrust into the abutments at either side. The earliest known arch bridges were built by the Greeks, and include the Arkadiko Bridge. With the span of 220 metres (720 ft), the Solkan Bridge over the Soèa River at Solkan in Slovenia is the second largest stone bridge in the world and the longest railroad stone bridge. It was completed in 1905. Its arch, which was constructed from over 5,000 tonnes (4,900 long tons; 5,500 short tons) of stone blocks in just 18 days, is the second largest stone arch in the world, surpassed only by the Friedensbrücke (Syratalviadukt) in Plauen, and the largest railroad stone arch.

The arch of the Friedensbrücke, which was built in the same year, has the span of 90 m (300 ft) and crosses the valley of the Syrabach River. The difference between the two is that the Solkan Bridge was built from stone blocks, whereas the Friedensbrücke was built from a mixture of crushed stone and cement mortar. The world's current largest arch bridge is the Chaotianmen Bridge over the Yangtze River with a length of 1,741 m (5,712 ft) and a span of 552 metres (1,811 ft). The bridge was open April 29, 2009 in Chongqing, China.Tied arch bridge Tied arch bridges have an arch-shaped superstructure, but differ from conventional arch bridges. Instead of transferring the weight of the bridge and traffic loads into thrust forces into the abutments, the ends of the arches are restrained by tension in the bottom chord of the structure. They are also called bowstring arches.

Suspension bridge Suspension bridges are suspended from cables. The earliest suspension bridges were made of ropes or vines covered with pieces of bamboo. In modern bridges, the cables hang from towers that are attached to caissons or cofferdams. The caissons or cofferdams are implanted deep into the floor of a lake or river. Subtypes include the simple suspension bridge, the stressed ribbon bridge, the underspanned suspension bridge, the suspended-deck suspension bridge, and the self-anchored suspension bridge. The longest suspension bridge in the world is the 3,909 m (12,825 ft) Akashi Kaikyô Bridge in Japan.

Cable-stayed bridgeCable-stayed bridges, like suspension bridges, are held up by cables. However, in a cable-stayed bridge, less cable is required and the towers holding the cables are proportionately higher. The first known cable-stayed bridge was designed in 1784 by C. T. (or C. J.) Löscher. The longest cable-stayed bridge is the Sutong Bridge over the Yangtze River in China.

Fixed or Movable Bridges

Figure: *A collapsed span of this bridge was reopened with a temporary bridge.*

Most bridges are fixed bridges, meaning they have no moving parts and stay in one place until they fail or are demolished. Temporary bridges, such as Bailey bridges, are designed to be assembled, and taken apart, transported to a different site, and re-used. They are important in military engineering, and are also used to carry traffic while an old bridge is being rebuilt. Movable bridges are designed to move out of the way of boats or other kinds of traffic, which would otherwise be too tall to fit. These are generally electrically powered.

Double-decked Bridges

Double-decked (or double-decker) bridges have two levels, such as the George Washington Bridge, connecting New York City to Bergen County, New Jersey, USA, as the world's busiest bridge, carrying 102 million vehicles annually; truss work between the roadway levels provided stiffness to the roadways and reduced movement of the upper level when the lower level was installed three decades following the upper level.

The Tsing Ma Bridge and Kap Shui Mun Bridge in Hong Kong have six lanes on their upper decks, and on their lower decks there are two lanes and a pair of tracks for MTR metro trains. Some double-decked bridges only use one level for street traffic; the Washington Avenue Bridge in Minneapolis reserves its lower level for automobile traffic and its upper level for pedestrian and bicycle traffic (predominantly students at the University of Minnesota). Likewise, in Toronto, the Prince Edward Viaduct has five lanes of motor traffic, bicycle lanes, and sidewalks on its upper deck; and a pair of tracks for the Bloor–Danforth subway line on its lower deck. The western span of the San Francisco – Oakland Bay Bridge also has two levels.

Robert Stephenson's High Level Bridge across the River Tyne in Newcastle upon Tyne, completed in 1849, is an early example of a double-decked bridge. The upper level carries a railway, and the lower level is used for road traffic. Other examples include Britannia Bridge over the Menai Strait and Craigavon Bridge in Derry, Northern Ireland. The Öresund Bridge between Copenhagen and Malmö consists of a four-lane highway on the upper level and a pair of railway tracks at the lower level. The Tower Bridge is different example of a double-decked bridge, with the central section consisting of a low level bascule span and a high level footbridge.

Viaducts

A viaduct is made up of multiple bridges connected into one longer structure. The longest and some of the highest bridges are viaducts, such as the Lake Pontchartrain Causeway and Millau Viaduct.

Figure: *A 21-arch bridge spanning Yorkshire's Wharfe valley, engineered for the Leeds and Thirsk Railway circa 1850.*

The longest viaduct in antiquity may have been the Pont Serme which crossed wide marshes in southern France. In Romance languages, the word *viaduct* refers to a bridge which spans only land. A bridge spanning water is called *ponte*.

Over Land

Viaducts are commonly used in many cities that are railroad centres, such as Chicago, Atlanta, Birmingham, London, and Manchester. These viaducts cross the large railroad yards that are needed for freight trains there, and also cross the multi-track railroad lines that are needed for heavy railroad traffic. These viaducts keep highway and city street traffic from having to be continually interrupted by the train traffic. Likewise, some viaducts carry railroads over large valleys, or they carry railroads over cities with many cross-streets and avenues.

Many viaducts over land connect points of similar height in a landscape, usually by bridging a river valley or other eroded opening in an otherwise flat area. Often such valleys had roads descending either side (with a small bridge over the river, where necessary) that become inadequate for the traffic load, necessitating a viaduct for "through" traffic. Such bridges also lend themselves for use by rail traffic, which requires straighter and flatter routes. Some viaducts have more than one deck, such that one deck has vehicular traffic and another deck having rail traffic. One example of this is the Prince

Edward Viaduct in Toronto, Canada, that carries motor traffic on the top deck as Bloor Street, and metro as the Bloor-Danforth subway line on the lower deck, over the steep Don River valley. Others were built to span settled areas and crossed over roads beneath - the reason for many viaducts in London.

Over Water

Viaducts over water are often combined with other types of bridges or tunnels to cross navigable waters. The viaduct sections, while less expensive to design and build than tunnels or bridges with larger spans, typically lack sufficient horizontal and vertical clearance for large ships.

The Millau Viaduct is a cable-stayed road-bridge that spans the valley of the River Tarn near Millau in southern France. Designed by the French bridge engineer Michel Virlogeux, in collaboration with architect Norman Robert Foster, it is the tallest vehicular bridge in the world, with one pier's summit at 343 metres (1,125 ft)—slightly taller than the Eiffel Tower and only 38 m (125 ft) shorter than the Empire State Building. It was formally dedicated on 14 December 2004 and opened to traffic two days later. The viaduct Danyang–Kunshan Grand Bridge in China is the longest bridge in the world according to Guinness World Records as of 2011.

Past and Future

Elevated expressways were built in rich cities such as Boston (Central Artery), Seoul, Tokyo, Toronto (Gardiner Expressway). Some were demolished because they were ugly and divided the city. However in developing nations such as Thailand, India (Delhi-Gurgaon Expressway), China, Bangladesh, Pakistan, elevated expressways have been built and more are under construction to improve traffic flow, particularly as a workaround of land shortage when built atop surface roads. In Indonesia, Viaduct system is usually used for Railway in Java also for Highway such as Jakarta Inner Ring Road.

Bridge Types by Use

A bridge can be categorized by what it is designed to carry, such as trains, pedestrian or road traffic, a pipeline or waterway for water transport or barge traffic. An aqueduct is a bridge that carries water, resembling a viaduct, which is a bridge that connects points of equal height. A road-rail bridge carries both road and rail traffic. A bridge can carry overhead power lines as does the Storstrøm Bridge.

Some bridges accommodate other purposes, such as the tower of Nový Most Bridge in Bratislava, which features a restaurant, or a bridge-restaurant which is a bridge built to serve as a restaurant. Other suspension bridge towers carry transmission antennas. Bridges are subject to unplanned uses as well. The areas underneath some bridges have become makeshift shelters and homes to homeless people, and the undersides of bridges all around the world are spots of prevalent graffiti. Some bridges attract people attempting suicide, and become known as suicide bridges.

Bridge Types by Material

Figure: *The Iron Bridge completed in 1779 was the first cast iron bridge.*

The materials used to build the structure are also used to categorize bridges. Until the end of the 18th Century, bridges were made out of timber, stone and masonry. Modern bridges are currently built in concrete, steel, fibre reinforced polymers (FRP), stainless steel or combinations of those materials.

Aesthetics

Most bridges are utilitarian in appearance, but in some cases, the appearance of the bridge can have great importance. Often, this is the case with a large bridge that serves as an entrance to a city, or crosses over a main harbor entrance. These are sometimes known as *signature bridges*. Designers of bridges in parks and along parkways often place more importance to aesthetics, as well. Examples include the stone-faced bridges along the Taconic State Parkway in New York.

To create a beautiful image, some bridges are built much taller than necessary. This type, often found in east-Asian style gardens, is called a Moon bridge, evoking a rising full moon. Other garden bridges may cross only a dry bed of stream washed pebbles, intended only to convey an impression of a stream. Often in palaces a bridge will be built over an artificial waterway as symbolic of a passage to an important place or state of mind. A set of five bridges cross a sinuous waterway in an important courtyard of the Forbidden City in Beijing, China. The central bridge was reserved exclusively for the use of the Emperor, Empress, and their attendants.

Bridge Maintenance

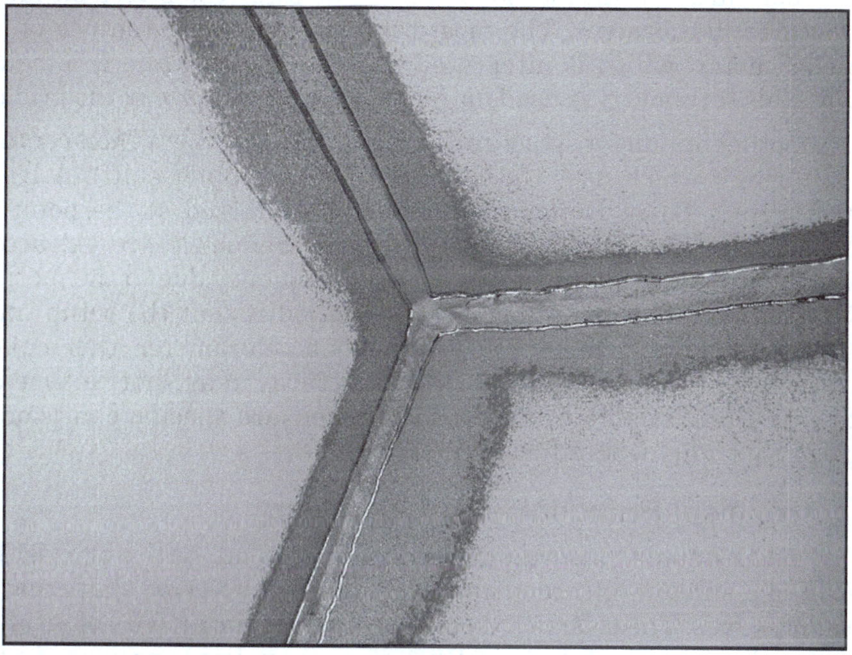

Figure: *Highway bridge with steel hollow box sections, performed for lifetime extension by welding aftertreatment*

Bridge maintenance consisting of a combination of structural health monitoring and testing. This is regulated in country-specific engineer standards and includes e.g. an ongoing monitoring every three to six months, a simple test or inspection every two to three years and a major inspection every six to ten years. In Europe, the cost of maintenance is higher than spending on new bridges. The lifetime of welded steel bridges can be significantly extended by aftertreatment of the weld transitions . This results in a potential high benefit, using existing bridges far beyond the planned lifetime.

Bridge Failures

The failure of bridges is of special concern for structural engineers in trying to learn lessons vital to bridge design, construction and maintenance. The failure of bridges first assumed national interest during the Victorian era when many new designs were being built, often using new materials. In the United States, the National Bridge Inventory tracks the structural evaluations of all bridges, including designations such as "structurally deficient" and "functionally obsolete".

Bridge Monitoring

There are several methods used to monitor the stress on large structures like bridges. The most common method is the use of an accelerometer, which is integrated into the bridge while it is being built. This technology is used for long-term surveillance of the bridge.

Another option for structural-integrity monitoring is "non-contact monitoring", which uses the Doppler effect (Doppler shift). A laser beam from a Laser Doppler Vibrometer is directed at the point of interest, and the vibration amplitude and frequency are extracted from the Doppler shift of the laser beam frequency due to the motion of the surface. The advantage of this method is that the setup time for the equipment is faster and, unlike an accelerometer, this makes measurements possible on multiple structures in as short a time as possible. Additionally, this method can measure specific points on a bridge that might be difficult to access.

Geotechnical Earthquake Considerations

An earthquake (also known as a quake, tremor or temblor) is the result of a sudden release of energy in the Earth's crust that creates seismic waves. The seismicity, seismism or seismic activity of an area refers to the frequency, type and size of earthquakes experienced over a period of time.

Earthquakes are measured using observations from seismometers. The moment magnitude is the most common scale on which earthquakes larger than approximately 5 are reported for the entire globe. The more numerous earthquakes smaller than magnitude 5 reported by national seismological observatories are measured mostly on the local magnitude scale, also referred to as the Richter scale. These two scales are numerically similar over their range of validity. Magnitude 3 or lower earthquakes are mostly almost imperceptible or weak and magnitude 7 and over potentially cause serious damage over larger areas, depending on their depth. The largest earthquakes in historic times have been of magnitude slightly over 9, although there is no limit to the possible magnitude. The most recent large earthquake of magnitude 9.0 or larger was a 9.0 magnitude earthquake in Japan in 2011 (as of October 2012), and it was the largest Japanese earthquake since records began. Intensity of shaking is measured on the modified Mercalli scale. The shallower an earthquake, the more damage to structures it causes, all else being equal.

At the Earth's surface, earthquakes manifest themselves by shaking and sometimes displacement of the ground. When the epicenter of a large earthquake is located offshore, the seabed may be displaced sufficiently to cause a tsunami. Earthquakes can also trigger landslides, and occasionally volcanic activity.

In its most general sense, the word *earthquake* is used to describe any seismic event — whether natural or caused by humans — that generates seismic waves. Earthquakes are caused mostly by rupture of geological faults, but also by other events such as volcanic activity, landslides, mine blasts, and nuclear tests. An earthquake's point of initial rupture is called its focus or hypocenter. The epicenter is the point at ground level directly above the hypocenter.

Bridges are big and beautiful structures, but they also need to be safe for the people who cross over them every day. Building a bridge that is safe and secure is a challenge to civil engineers.

Earthquake prone areas always exert great pressure on economy and social security. Safety is the most important aspect of bridge building for earthquake areas. Modern technology has now enabled engineers to build earthquake resistant bridges economically yet keeping the reliability factor intact.

Bridges do not just connect two far-off places; they serve a greater purpose. They minimize distance and help humans win over natural obstructions, rivers in particular. Building earthquake resistant *buildings* is important for human safety, but if a *bridge* is destroyed, connected places can be completely isolated from each other.

The 1989 San Francisco Bay Area earthquake suspended traffic for more than twenty hours because of the damage done to a bridge deck. In 2011, Japan suffered a major earthquake resulting in many bridge collapses, but the losses were minimized because the Japanese implemented earthquake resistant design practices. There have been many incidents in the US and other countries where post-earthquake studies have reported that a seismic design would have minimized the losses- or even absorbed the earthquake jolts.

The Design Considerations

While constructing bridges in earthquake prone areas, especially in the seismically active Zones III and IV of the Indian subcontinent, one thing we know in advance is that the chances of earthquakes occurring in such areas are very high. And that is the only thing we know because predicting earthquakes is not yet possible.

While designing bridges in earthquake prone areas following things must be considered.

Earthquake history of the region is a major factor that needs due attention from engineers. If a particular seismic zone experiences earthquakes of different magnitudes periodically, historical facts will help in predicting the most probable times of the year when an earthquake can happen. Obviously, you do not want to construct a bridge at the time when the chances of earthquake are high. Bridge building for earthquake areas needs extensive study of the earthquake history of the region.

Load Factors: Only dead and live loads are considered when designing bridges in non-earthquake zones. However, in quake prone areas, another force is of concern. This is called "seismic load," and it includes the forces on the bridge due to acceleration produced by earthquake jolts. How do we fit seismic load in our design? Historical facts help us to determine the seismic load factor in our designs. In case a region has no earthquake history, a minimum load is considered in design, which varies from zone to zone.

Seismic retrofitting is what modern technology has to offer. It not only makes old bridges resistant to earthquakes, but also helps in cost reduction. (Constructing a new bridge is always a time and resource consuming practice.) After a detailed seismic performance evaluation, retrofitting is carried out on old or weak bridges using methods such as friction damper systems, carbon fibre plastic reinforcements, and external prestressing, as well as improving soil properties to keep a check on ground motion. Seismic retrofitting is a relatively new technique but the Civil Engineering societies and associations across the world are promoting it because this is going to save money, time, and resources, without compromising the reliability and safety factor.

Minimizing Risk - The Bottomline

Japan, the pioneer of earthquake engineering, lost many lives and suffered huge property loss in 2011. Mastering technology is something that we humans can do, but the bottom line is that we cannot beat nature. This does not mean we should be pessimistic or passive; it means that we need to put our best foot forward and try to minimize the risk. Earthquakes won't affect us as much if we are prepared, and the best preparation at this time is to implement earthquake resistant design practices and to make the best use of seismic retrofitting technology. An optimistic approach and staying updated with information related to seismic zones and seismic design considerations will definitely help to build better and more stable bridges.

Seismic Analysis

Seismic analysis is a subset of structural analysis and is the calculation of the response of a building (or nonbuilding) structure to earthquakes. It is part of the process of structural design, earthquake engineering or structural assessment and retrofit in regions where earthquakes are prevalent.

A building has the potential to 'wave' back and forth during an earthquake (or even a severe wind storm). This is called the 'fundamental mode', and is the lowest frequency of building response. Most buildings, however, have higher modes of response, which are uniquely activated during earthquakes. The second mode, but there are higher 'shimmy' (abnormal vibration) modes. Nevertheless, the first and second modes tend to cause the most damage in most cases.

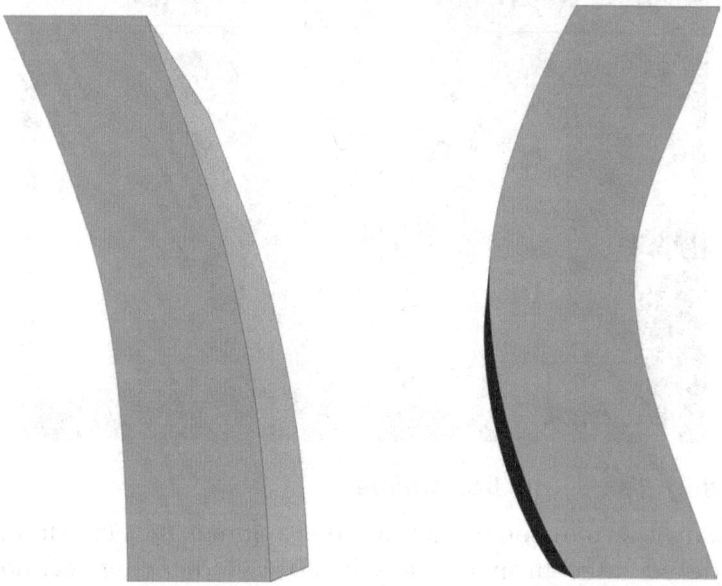

Figure: *First and second modes of building seismic response*

The earliest provisions for seismic resistance were the requirement to design for a lateral force equal to a proportion of the building weight (applied at each floor level). This approach was adopted in the appendix of the 1927 Uniform Building Code (UBC), which was used on the west coast of the USA. It later became clear that the dynamic properties of the structure affected the loads generated during an earthquake.

In the Los Angeles County Building Code of 1943 a provision to vary the load based on the number of floor levels was adopted (based on research carried out at Caltech in collaboration with Stanford University and the U.S. Coast and Geodetic Survey, which started in 1937). The concept of "response spectra" was developed in the 1930s, but it wasn't until 1952 that a joint committee of the San Francisco Section of the ASCE and the Structural Engineers Association of Northern California (SEAONC) proposed using the building period (the inverse of the frequency) to determine lateral forces.

The University of California, Berkeley was an early base for computer-based seismic analysis of structures, led by Professor Ray Clough (who coined the term finite element). Students included Ed Wilson, who went on to write the programme SAP in 1970, an early "Finite Element Analysis" programme.

Earthquake engineering has developed a lot since the early days, and some of the more complex designs now use special earthquake protective elements either just in the foundation (base isolation) or distributed throughout the structure. Analyzing these types of structures requires specialized explicit finite element computer code, which divides time into very small slices and models the actual physics, much like common video games often have "physics engines". Very large and complex buildings can be modelled in this way (such as the Osaka International Convention Centre). Structural analysis methods can be divided into the following five categories.

Equivalent Static Analysis

This approach defines a series of forces acting on a building to represent the effect of earthquake ground motion, typically defined by a seismic design response spectrum. It assumes that the building responds in its fundamental mode. For this to be true, the building must be low-rise and must not twist significantly when the ground moves. The response is read from a design response spectrum, given the natural frequency of the building (either calculated or defined by the building code). The applicability of this method is extended in many building codes by applying factors to account for higher buildings with some higher modes, and for low levels of twisting. To account for effects due to "yielding" of the structure, many codes apply modification factors that reduce the design forces (e.g. force reduction factors).

Response Spectrum

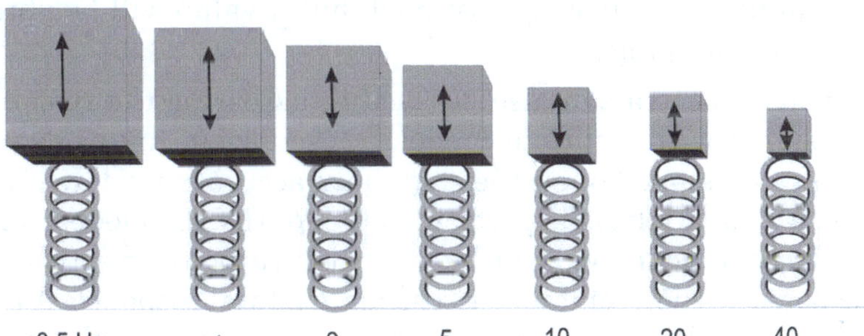

Figure: *A series of mixed vertical oscillators*

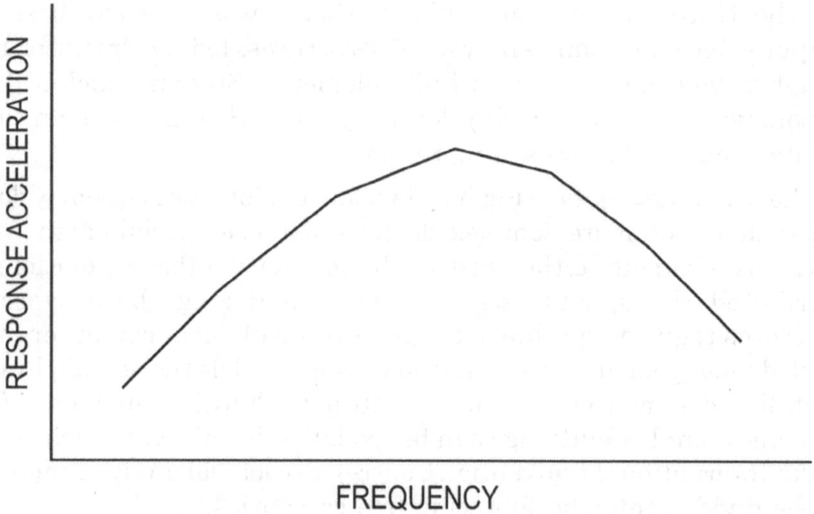

Figure: *A plot of the peak acceleration for the mixed vertical oscillators*

A response spectrum is simply a plot of the peak or steady-state response (displacement, velocity or acceleration) of a series of oscillators of varying natural frequency, that are forced into motion by the same base vibration or shock. The resulting plot can then be used to pick off the response of any linear system, given its natural frequency of oscillation. One such use is in assessing the peak response of buildings to earthquakes. The science of strong ground motion may use some values from the ground response spectrum (calculated from recordings of surface ground motion from seismographs) for correlation with seismic damage.

If the input used in calculating a response spectrum is steady-state periodic, then the steady-state result is recorded. Damping must be present, or else the response will be infinite. For transient input (such as seismic ground motion), the peak response is reported. Some level of damping is generally assumed, but a value will be obtained even with no damping.

Response spectra can also be used in assessing the response of linear systems with multiple modes of oscillation (multi-degree of freedom systems), although they are only accurate for low levels of damping. Modal analysis is performed to identify the modes, and the response in that mode can be picked from the response spectrum. This peak response is then combined to estimate a total response. A typical combination method is the square root of the sum of the squares (SRSS) if the modal frequencies are not close. The result is typically different from that which would be calculated directly from an input, since phase information is lost in the process of generating the response spectrum.

The main limitation of response spectra is that they are only universally applicable for linear systems. Response spectra can be generated for non-linear systems, but are only applicable to systems with the same non-linearity, although attempts have been made to develop non-linear seismic design spectra with wider structural application. The results of this cannot be directly combined for multi-mode response.

Seismic Response Spectra

Response spectra are very useful tools of earthquake engineering for analyzing the performance of structures and equipment in earthquakes, since many behave principally as simple oscillators (also known as single degree of freedom systems). Thus, if you can find out the natural frequency of the structure, then the peak response of the building can be estimated by reading the value from the ground response spectrum for the appropriate frequency. In most building codes in seismic regions, this value forms the basis for calculating the forces that a structure must be designed to resist (seismic analysis).

As mentioned earlier, the ground response spectrum is the response plot done at the free surface of the earth. Significant seismic damage may occur if the building response is 'in tune' with components of the ground motion (resonance), which may be identified from the response spectrum. This was observed in the 1985 Mexico City Earthquake where the oscillation of the deep-soil lake bed was similar to the natural frequency of mid-rise concrete buildings, causing significant damage. Shorter (stiffer) and taller (more flexible) buildings suffered less damage.

In 1941 at Caltech, George W. Housner began to publish calculations of response spectra from accelerographs. In the 1982 EERI Monograph on "Earthquake Design and Spectra", Newmark and Hall describe how they developed an "idealized" seismic response spectrum based on a range of response spectra generated for available earthquake records. This was then further developed into a design response spectrum for use in structural design, and this basic form (with some modifications) is now the basis for structural design in seismic regions throughout the world (typically plotted against structural "period", the inverse of frequency). A nominal level of damping is assumed (5% of critical damping).

For "regular" low-rise buildings, the structural response to earthquakes is characterized by the fundamental mode (a "waving" back-and-forth), and most building codes permit design forces to be calculated from the design spectrum on the basis of that frequency, but for more complex structures, combination of the results for many modes (calculated through modal analysis) is often required. In extreme cases, where structures are either too irregular, too tall or of significance to a community in disaster response, the response spectrum approach is no longer appropriate, and more complex analysis is required, such as non-linear static or dynamic analysis like in seismic performance analysis technique.

Response Spectrum Analysis

This approach permits the multiple modes of response of a building to be taken into account (in the frequency domain). This is required in many building codes for all except for very simple or very complex structures. The response of a structure can be defined as a combination of many special shapes (modes) that in a vibrating string correspond to the "harmonics". Computer analysis can be used to determine these modes for a structure.

For each mode, a response is read from the design spectrum, based on the modal frequency and the modal mass, and they are then combined to provide an estimate of the total response of the structure. In this we have to calculate the magnitude of forces in all directions i.e. X, Y & Z and then see the effects on the building.. Combination methods include the following:

- absolute - peak values are added together
- square root of the sum of the squares (SRSS)
- complete quadratic combination (CQC) - a method that is an improvement on SRSS for closely spaced modes

The result of a response spectrum analysis using the response spectrum from a ground motion is typically different from that which would be calculated directly from a linear dynamic analysis using that ground motion directly, since phase information is lost in the process of generating the response spectrum.

In cases where structures are either too irregular, too tall or of significance to a community in disaster response, the response spectrum approach is no longer appropriate, and more complex analysis is often required, such as non-linear static analysis or dynamic analysis.

Linear Dynamic Analysis

Static procedures are appropriate when higher mode effects are not significant. This is generally true for short, regular buildings. Therefore, for tall buildings, buildings with torsional irregularities, or non-orthogonal systems, a dynamic procedure is required. In the linear dynamic procedure, the building is modelled as a multi-degree-of-freedom (MDOF) system with a linear elastic stiffness matrix and an equivalent viscous damping matrix.

The seismic input is modelled using either modal spectral analysis or time history analysis but in both cases, the corresponding internal forces and displacements are determined using linear elastic analysis. The advantage of these linear dynamic procedures with respect to linear static procedures is that higher modes can be considered. However, they are based on linear elastic response and hence the applicability decreases with increasing nonlinear behaviour, which is approximated by global force reduction factors.

In linear dynamic analysis, the response of the structure to ground motion is calculated in the time domain, and all phase information is therefore maintained. Only linear properties are assumed. The analytical method can use modal decomposition as a means of reducing the degrees of freedom in the analysis.

Nonlinear Static Analysis

In general, linear procedures are applicable when the structure is expected to remain nearly elastic for the level of ground motion or when the design results in nearly uniform distribution of nonlinear response throughout the structure. As the performance objective of the structure implies greater inelastic demands, the uncertainty with linear procedures increases to a point that requires a high level of conservatism in demand assumptions and acceptability criteria to avoid unintended performance. Therefore, procedures incorporating inelastic analysis can reduce the uncertainty and conservatism.

This approach is also known as "pushover" analysis. A pattern of forces is applied to a structural model that includes non-linear properties (such as steel yield), and the total force is plotted against a reference displacement to define a capacity curve. This can then be combined with a demand curve (typically in the form of an acceleration-displacement response spectrum (ADRS)). This essentially reduces the problem to a single degree of freedom (SDOF) system.

Nonlinear static procedures use equivalent SDOF structural models and represent seismic ground motion with response spectra. Story drifts and component actions are related subsequently to the global demand parameter by the pushover or capacity curves that are the basis of the non-linear static procedures.

Nonlinear Dynamic Analysis

Nonlinear dynamic analysis utilizes the combination of ground motion records with a detailed structural model, therefore is capable of producing results with relatively low uncertainty. In nonlinear dynamic analyses, the detailed structural model subjected to a ground-motion record produces estimates of component deformations for each degree of freedom in the model and the modal responses are combined using schemes such as the square-root-sum-of-squares.

In non-linear dynamic analysis, the non-linear properties of the structure are considered as part of a time domain analysis. This approach is the most rigorous, and is required by some building codes for buildings of unusual configuration or of special importance. However, the calculated response can be very sensitive to the characteristics of the individual ground motion used as seismic input; therefore, several analyses are required using different ground motion records to achieve a reliable estimation of the probabilistic distribution of structural response. Since the properties of the seismic response depend on the intensity, or severity, of the seismic shaking, a comprehensive assessment calls for numerous nonlinear dynamic analyses at various levels of intensity to represent different possible earthquake scenarios. This has led to the emergence of methods like the Incremental Dynamic Analysis.

Peak Ground Acceleration

Peak ground acceleration (PGA) is a measure of earthquake acceleration on the ground and an important input parameter for earthquake engineering, also known as the design basis earthquake ground motion (DBEGM) Unlike the Richter and moment magnitude scales, it is not a measure of the total energy (magnitude, or size) of an earthquake, but rather of how hard the earth shakes in a given geographic area (the *intensity*). The Mercalli intensity scale uses personal reports and observations to measure earthquake intensity but PGA is measured by instruments, such as accelerographs, and it generally correlates well with the Mercalli scale.

The peak horizontal acceleration (PHA) is the most commonly used type of ground acceleration in engineering applications, and is used to set building codes and design hazard risks. In an earthquake, damage to buildings and infrastructure is related more closely to ground motion, rather than the magnitude of the earthquake. For moderate earthquakes, PGA is the best determinate of damage; in severe earthquakes, damage is more often correlated with peak ground velocity.

Geophysics

Earthquake energy is dispersed in waves from the epicentre, causing ground movement horizontally (in two directions) and vertically. PGA records the acceleration (rate of change of speed) of these movements, while peak ground velocity is the greatest speed (rate of movement) reached by the ground, and peak displacement is the distance moved. These values vary in different earthquakes, and in differing sites within one earthquake event, depending on a number of factors. These include the length of the fault, magnitude, the depth of the quake, the distance from the epicentre, the duration (length of the shake cycle), and the geology of the ground (subsurface). Shallow-focused earthquakes generate stronger shaking (acceleration) than intermediate and deep quakes, since the energy is released closer to the surface.

Peak ground acceleration can be expressed in *g* (the acceleration due to Earth's gravity, equivalent to g-force) as either a decimal or percentage; in m/s^2 ($1\,g = 9.81\ m/s^2$); or in Gal, where 1 Gal is equal to $0.01\ m/s^2$ ($1\,g = 981$ Gal).

The ground type can significantly influence ground acceleration, so PGA values can display extreme variability over distances of a few kilometers, particularly with moderate to large earthquakes. The varying PGA results from an earthquake can be displayed on a shake map. Due to the complex conditions affecting PGA, earthquakes of similar magnitude can offer disparate results, with many moderate magnitude earthquakes generating significantly larger PGA values than larger magnitude quakes.

During an Earthquake, Ground Acceleration is Measured in Three Directions: vertically (V or UD, for up-down) and two perpendicular horizontal directions (H1 and H2), often north-south (NS) and east-west (EW). The peak acceleration in each of these directions is recorded, with the highest individual value often reported.

Alternatively, a combined value for a given station can be noted. The peak horizontal ground acceleration (PHA or PHGA) can reached by selecting the higher individual recording, taking the mean of the two values, or calculating a vector sum of the two components.

A three-component value can also be reached, by taking the vertical component into consideration also.

In seismic engineering, the effective peak acceleration (EPA, the maximum ground acceleration to which a building responds) is often used, which tends to be ⅔ – ¾ the PGA.

Seismic Risk and Engineering

Study of geographic areas combined with an assessment of historical earthquakes allows geologists to determine seismic risk and to create seismic hazard maps, which show the likely PGA values to be experienced in a region during an earthquake, with a probability of exceedance (PE).

Seismic engineers and government planning departments use these values to determine the appropriate earthquake loading for buildings in each zone, with key identified structures (such as hospitals, bridges, power plants) needing to survive the maximum considered event (MCE).

Damage to buildings is related to both peak ground velocity and PGA, and the duration of the earthquake – the longer high-level shaking persists, the greater the likelihood of damage.

Comparison of Instrumental and Felt Intensity

Peak ground acceleration provides a measurement of *instrumental intensity*, that is, ground shaking recorded by seismic instruments. Other intensity scales measure *felt intensity*, based on eyewitness reports, felt shaking, and observed damage. There is correlation between these scales, but not always absolute agreement since experiences and damage can be affected by many other factors, including the quality of earthquake engineering.

Generally speaking,

- 0.001 g (0.01 m/s^2) – perceptible by people
- 0.02 g (0.2 m/s^2) – people lose their balance
- 0.50 g – very high; well-designed buildings can survive if the duration is short.

Correlation with the Mercalli Scale

The United States Geological Survey developed an Instrumental Intensity scale which maps peak ground acceleration and peak ground velocity on an intensity scale similar to the felt Mercalli scale. These values are used to create shake maps by seismologists around the world.

Instrumental Intensity	*Acceleration (g)*	*Velocity (cm/s)*	*Perceived Shaking*	*Potential Damage*
I	< 0.0017	< 0.1	Not felt	None
II-III	0.0017 - 0.014	0.1 - 1.1	Weak	None
IV	0.014 - 0.039	1.1 - 3.4	Light	None
V	0.039 - 0.092	3.4 - 8.1	Moderate	Very light
VI	0.092 - 0.18	8.1 - 16	Strong	Light
VII	0.18 - 0.34	16 - 31	Very strong	Moderate
VIII	0.34 - 0.65	31 - 60	Severe	Moderate to heavy
IX	0.65 - 1.24	60 - 116	Violent	Heavy
X+	> 1.24	> 116	Extreme	Very heavy

Other Intensity Scales

In the 7-class Japan Meteorological Agency seismic intensity scale, the highest intensity, Shindo 7, covers accelerations greater than 4 m/s^2 (0.41 *g*).

The Japan Meteorological Agency seismic intensity scale is a seismic scale used in Japan and Taiwan to measure the intensity of earthquakes. It is measured in units of *shindo* (seismic intensity, lit. "degree of shaking").

Unlike the moment magnitude scale (formerly Richter scale), which measures the energy released by the earthquake, the JMA scale describes the degree of shaking at a point on the Earth's surface, and is analogous to the Mercalli intensity scale.

The intensity of an earthquake is not totally determined by its magnitude, and varies from place to place; for example, a quake may be described as "shindo 4 in Tokyo, shindo 3 in Yokohama, shindo 2 in Shizuoka".

The JMA operates a network of 180 seismographs and 627 seismic intensity metres and provides real-time earthquake reports to the media and on the internet.

PGA Hazard Risks Worldwide

In India, areas with expected PGA values higher than 0.36*g* are classed as "Zone 5", or "Very High Damage Risk Zone".

Earthquake Damage to Bridges

Earthquake damage to a bridge can have severe consequences. Clearly, the collapse of a bridge places people on or below the bridge at risk, and it must be replaced after the earthquake unless alternative transportation paths are identified. The consequences of less severe damage are less obvious and dramatic, but they are nonetheless important. A bridge closure, even if it is temporary, can have tremendous consequences, because bridges often provide vital links in a transportation system. In the immediate aftermath of an earthquake, closure of a bridge can impair emergency response operations. Later, the economic impact of a bridge closure increases with the length of time the bridge is closed, the economic importance of the traffic using the route, the traffic delay caused by following alternate routes, and the replacement cost for the bridge.

Even when the cause of a particular collapse is well understood, it is difficult to generalize about the causes of bridge damage. In past earthquakes, the nature and extent of damage that each bridge suffered have varied with the characteristics of the ground motion at the particular site and the construction details of the particular bridge. No two earthquakes or bridge sites are identical. Design and construction practices vary extensively throughout the world and even within the United States. These practices have evolved with time, and, in particular, seismic design practice improved significantly in the western United States during the 1970s as a result of experience gained from the 1971

San Fernando Arthquake

Despite these uncertainties and variations, one can learn from past earthquake damage, because many types of damage occur repeatedly. By being aware of typical vulnerabilities that bridges have experienced, it is possible to gain insight into structural behaviour and to identify potential weaknesses in existing and new bridges. Historically, observed damage has provided the impetus for many improvements in earthquake engineering codes and practice.

An effort is made to distinguish damage according to two classes, as follows:

Primary damage — Damage caused by earthquake ground shaking or deformation that was the primary cause of damage to the bridge, and that may have triggered other damage or collapse. *Secondary damage* — Damage caused by earthquake ground shaking or deformation that was the result of structural failures elsewhere in the bridge, and was caused by redistribution of internal actions for which the structure was not designed.

The emphasis is on primary damage. It must be accepted, however, that in many cases the distinction between primary and secondary damage is obscure because the bridge geometry is complex or, in the case of collapse, because it is difficult to reconstruct the failure sequence.

Effects of Site Conditions

Performance of a bridge structure during an earthquake is likely to be influenced by proximity of the bridge to the fault and site conditions. Both of these factors affect the intensity of ground shaking and ground deformations, as well as the variability of those effects along the length of the bridge.

The influence of site conditions on bridge response became widely recognized following the 1989 Loma Prieta earthquake. Plots the locations of minor and major bridge damage from the Loma Prieta earthquake. With some exceptions, the most significant damage occurred around the perimetre or within San Francisco Bay where relatively deep and soft soil deposits amplified the bedrock ground motion.

In the same earthquake, the locations of collapse of the Cypress Street Viaduct nearly coincided with zones of natural and artificial fill where ground shaking was likely to have been the strongest. A major conclusion to be drawn from this and other earthquakes is the significant impact that local site conditions have on amplifying strong ground motion, and the subsequent increased vulnerability of bridges on soft soil sites. This observation is important because many bridges and elevated roadways traverse bodies of water where soft soil deposits are common.

During the 1995 Hyogo-Ken Nanbu (Kobe) earthquake, significant damage and collapse likewise occurred in elevated roadways and bridges founded adjacent to or within Osaka Bay. Several types of site conditions contributed to the failures. First, many of the bridges were founded on sand–gravel terraces (alluvial deposits) overlying gravel–sand–mud deposits at depths of less than 33 ft (10 m), a condition which is believed to have led to site amplification of the bedrock motions.

Furthermore, many of the sites were subject to liquefaction and lateral spreading, resulting in permanent substructure deformations and loss of superstructure support. Finally, the site was directly above the fault rupture, resulting in ground motions having high horizontal and vertical ground accelerations as well as large velocity pulses. Near-fault ground motions can impose large deformation demands on yielding structures, as was evident in the overturning collapse of all 17 bents of the Higashi-Nada Viaduct of the Hanshin Expressway, Route 3, in Kobe.

Correlation of Damage with Construction Era

Bridge seismic design practices have changed over the years, largely reflecting lessons learned from performance in past earthquakes. Several examples in the literature demonstrate that the constructionera of a bridge is a good indicator of likely performance, with higher damage levels expected in older construction than in newer construction.

An excellent example of the effect of construction era is provided by observing the relative performances of bridges on Routes 3 and 5 of the Hanshin Expressway in Kobe. Route 3 was constructed from 1965 through 1970, while Route 5 was completed in the early to mid-1990s. The two routes are parallel to one another, with Route 3 being farther inland and Route 5 being built largely on reclaimed land. Despite the potentially worse soil conditions for Route 5, it performed far better than Route 3, losing only a single span owing apparently to permanent ground deformation and span unseating. In contrast, Route 3 has been estimated to have sustained moderate-to-large-scale damage in 637 piers, with damage in over 1300 spans, and approximately 50 spans requiring replacement.

The superior performance of newer construction in the Hyogo-Ken Nanbu earthquake and other earthquakes has led to the use of benchmark years as a crude but effective method for rapidly assessing the likely performance of bridge construction. This method has been an effective tool for bridge assessment in California. The reason for its success there is the rapid change in bridge construction practice following the 1971 San Fernando earthquake. Before that time, California design and construction practice was based on significantly lower design forces and less stringent detailing requirements compared with current requirements. In the period following that earth- quake, the California Department of Transportation (Caltrans) developed new design approaches requiring increased strength and improved detailing for ductile response.

The 1994 Northridge earthquake provides an insightful study on the use of benchmarking. Over 2500 bridges existed in the metropolitan Los Angeles freeway system at that time.

Effects of Changes in Condition

Changes in the condition of a bridge can greatly affect its seismic performance. In many regions of

North America, extensive deterioration of bridge superstructures, bearings, and substructures has Construction modifications, either during the original construction or during the service life, can also have a major effect on bridge performance. Several graphic examples were provided by the Northridge earthquake. A bridge column that was unintentionally restrained by a reinforced concrete channel wall. The wall shortened the effective length of the column, increased the column shear force, and shifted nonlinear response from a zone of heavy confinement upward to a zone of light transverse reinforcement, where the ductility capacity was inadequate. Failures of this type illustrate the importance of careful inspection during construction and during the service life of a bridge.

Effects of Structural Configuration

Ideally, earthquake-resistant construction should be designed to have a regular configuration so that the behaviour is simple to conceptualize and analyze, and so that inelastic energy dissipation is promoted in a large number of readily identified yielding components. This ideal often is not achievable in bridge construction because of irregularities imposed by site conditions and traffic flow requirements. In theory, any member or joint can be configured to resist the induced force and deformation demands. However, in practice, bridges with certain configurations are more vulnerable to earthquakes than others.

Experience indicates that a bridge is most likely to be vulnerable if

(1) excessive deformation demands occur in a few brittle elements,

(2) the structural configuration is complex, or

(3) a bridge lacks redundancy.

The bridge designer needs to recognize the potential consequences of these irregularities and to design accordingly either to reduce the irregularity or to toughen the structure to compensate for it.

A common form of irregularity arises when a bridge traverses a basin requiring columns of nonuniform length. Although the response of the superstructure may be relatively uniform, the deformation demands on the individual substructure piers are highly irregular; the largest strains are imposed on the shortest columns. In some cases, the deformation demands on the short columns can induce their failure before longer, more flexible adjacent columns can fully participate. The Route 14/5 Separation and Overhead structure provides an example of these phenomena. The structure comprised a box-girder monolithic with single-column bents that varied in height depending on the road and grade elevations. Apparently, the short column at Bent 2 failed in shear because of large deformation demands in that column, resulting in the collapse of the adjacent spans.

The effects identified above can be exacerbated in long-span bridges. In addition to changes in subgrade and structural irregularities that may be required to resolve complex foundation and transportation requirements, long bridges can be affected by spatial and temporal variations in the ground motions. Expressed in simple terms, different piers are subjected to different ground motions at any one time, because seismic waves take time to travel from one bridge pier to another. This effect can result in one pier being pulled in one direction while the other is being pushed in the opposite direction. This complex behaviour is not accounted for directly in conventional bridge design. An example where this behaviour may have resulted in increased damage and collapse is the eastern portion of the San Francisco–Oakland Bay Bridge. This bridge includes a variety of different superstructure and substructure configurations, traverses variable subsoils, and is long enough for spatial and temporal variations in ground motions to induce large relative displacements between adjacent bridge segments. The bridge lost two spans, one upper and one lower, at a location where the superstructure was required to accommodate differential movements of adjacent bridge segments.

Unseating at Expansion Joints

Expansion joints introduce a structural irregularity that can have catastrophic consequences. Such joints are commonly provided in bridges to alleviate stresses associated with volume changes that occur as a bridge ages and as the temperature changes. These joints can occur within a span (in- span hinges), or they can occur at the supports, as is the case for simply supported bridges. Earthquake ground shaking, or transient or permanent ground deformations resulting from the earthquake, can induce superstructure movements that cause the supported span to unseat. Unseating is especially a problem with the shorter seats that were common in older construction (e.g., References).

Bridges with Short Seats and Simple Spans

In much of the United States and in many other areas of the world, bridges often comprise a series of simple spans supported on bents. These spans are prone to being toppled from their supporting substructures either due to shaking or differential support movement associated with ground deformation.

Unseating of simple spans was observed in California in earlier earthquakes, leading in recent decades to development of bridge construction practices based on monolithic box-girder- substructure construction. Problems of unseating still occur with older bridge construction and with new bridges in regions where simple spans are still common. For example, during the 1991

Costa Rica earthquake, widespread liquefaction led to abutment and internal bent rotations, resulting in the collapse of no fewer than four bridges with simple supports. The collapse of the Showa Bridge in the 1964 Niigata earthquake demonstrates one result of the unseating of simple spans

Skewed Bridges

Skewed bridges are defined as those having supports that are not perpendicular to the alignment of the bridge. Collisions between a skewed bridge and its abutments (or adjacent frames) can cause a bridge to rotate about a vertical axis. Because the abutments resist compression but not tension, the sense of this rotation is the same (for a given bridge configuration) regardless of whether thebridge collides with one abutment or the other. If the rotations are large and the seat lengths small, a bridge can come unseated at the acute corners of the decks.

Several examples of skewed bridge damage and collapse can be found in the literature. A typical example is the Rio Bananito Bridge, in which the bridge and central slab pier were skewed at 30°, which lost both spans off the central pier in the direction of the skew during the 1991 Costa Rica earthquake. Another example of skewed bridge failure is the Gavin Canyon Undercrossing, which failed during the 1994 Northridge earthquake. Both skewed hinges became unseated during the earthquake, resulting in collapse of the unseated spans.

Curved Bridges

Curved bridges can have asymmetrical response similar to that of skewed bridges. For loading in one direction, an in-span hinge tends to close, while for loading in the other direction, the hinge opens. An example in which the curved alignment may have contributed to bridge collapse is the curved ramps of the I-5/SR14 interchange, which sustained collapses in both the 1971 San Fernando earthquake and the 1994 Northridge earthquake. Other factors that may have contributed to the failures include inadequate hinge seats and column deformability.

Hinge Restrainers

Hinge restrainers appear to have been effective in preventing unseating in both the Loma Prieta and Northridge earthquakes. In some other cases, hinge restrainers were not fully effective in preventing unseating. For example, the hinge restrainers in the Gavin Canyon Undercrossing, which were aligned parallel the bridge alignment, did not prevent unseating.

Damage to Superstructures

Superstructures are designed to support service gravity loads elastically, and, for seismic applications, they are usually designed to be a strong link in the earthquake-resisting system. As a result, superstructures tend to be sufficiently strong to remain essentially elastic during earthquakes. In general, superstructure damage is unlikely to be the primary cause of collapse of a span.

Instead, damage typically is focused in bearings and substructures. The superstructure may rest on elastomeric pads, pin supports, or rocker bearings, or may be monolithic with the substructure. As bearings and substructures are damaged and in some cases collapse, a wide range of damage and failure of superstructures may result, but these failures are often secondary; that is, they result from failures elsewhere in the bridge. There are, however, some cases of primary superstructure damage as well. Some examples are highlighted below.

With the exception of bridge superstructures that come unseated and collapse, the most common form of damage to superstructures is due to pounding of adjacent segments at the expansion hinges. This type of damage occurs in bridges of all construction materials. Following the 1971 San Fernando earthquake, Caltrans initiated the first phase of its retrofit programme, which involved installation of hinge and joint restrainers to prevent deck joints from separating. Both cable restrainers and pipe restrainers (the former intended only to restrain longitudinal movement and the latter intended also to restrain transverse motions) were installed in bridge superstructures.

The restrainers extended through end diaphragms that had not been designed originally for the forces associated with restraint. Some punching shear damage to end diaphragms retrofitted with cable restrainers was observed in the I-580/I-980/SR24 connectors following the 1989 Loma Prietaearthquak. Steel superstructures commonly comprise lighter framing elements, especially for transverse bracing. These have been found to be susceptible to damage due to transverse loading, especially following failure of bearings. Several cases of steel superstructure damage occurred in the Hyogo-Ken Nanbu earthquake. The movement was sufficient for one end of one arch to drop off the cap beam, twisting the superstructure and apparently resulting in the buckling of the top chord bracing members.

A spectacular example of steel superstructure failure and collapse is that of the eastern portion of the San Francisco–Oakland Bay Bridge during the 1989 Loma Prieta earthquake. In this bridge, a 50-ft (15-m) span over tower E9 was a transition point between 506-ft (154-m) truss spans to the west and 290-ft (88-m) truss spans to the east, serving to transmit longitudinal forces among the adjacent spans and the massive steel tower at E9. Failure of a bolted connection between the 290-ft (88-m) span truss and the tower resulted in sliding of the span and unseating of the transition span over tower E9. This collapse resulted in closure for 1 month of this critical link between San Francisco and the East Bay.

Damage to Bearings

In some regions of the world, the prevalent bridge construction consists of steel superstructures supported on bearings, which, in turn, rest on a substructure. In the United States, this form of construction is common in new bridges east of the Sierra Nevada Mountains as well as throughout the country in older existing bridges. In such bridges, the bearings commonly consist of steel components designed to provide restraint in one or more directions and, in some cases, to permit movement in one or more directions. Failure of these bearings in an earthquake can cause redistribution of internal forces, which may overload either the superstructure or substructure, or both. Collapse is also possible when bearing support is lost.

The predominant type of bridge construction in Japan involves steel superstructures supported on bearings, which, in turn, are supported on concrete substructures. The Hyogo-Ken Nanbu earthquake provides several examples of bearing failures in these types of bridges. One example is provided by the Hamate Bypass, which was a double-deck elevated roadway comprising steel box girders on either fixed or expansion steel bearings. Another example is provided by the Nishinomiya-ko Bridge, a 830-ft (252-m) span-arch bridge supported on two fixed bearings at one end and two expansion bearings at the other end. The fixed-end bearings, which apparently were designed to have a capacity of approximately 70% of the bridge weight, failed, apparently leading to unseating of the adjacent approach span. The failed bearing is shown in

Damage to Substructures

Columns

Unlike building design, current practice in bridge design is to proportion members of a frame (bent) such that its lateral-load capacity is limited by the flexural strength of its columns. For this strategy to be successful, the connecting elements (e.g., footings, joints, cross-beams) need to be strong enough to force yielding into the columns, and the columns need to be sufficiently ductile (or tough) to sustain the imposed deformations. Even in older bridges, where the "weak column" design approach may not have been adopted explicitly, columns tend to be weaker than the beam–dia- phragm–slab assembly to which they connect. Consequently, columns can be subjected to large inelastic demands during strong earthquakes. Failure of a column can result in loss of vertical load- carrying capacity; column failure is often the primary cause of bridge collapse.

Most damage to columns can be attributed to inadequate detailing, which limits the ability of the column to deform inelastically. In concrete columns, the detailing inadequacies can produce flexural, shear, splice, or anchorage failures, or as is often the case, a failure that combines several mechanisms. In steel columns, local buckling has been observed to lead progressively to collapse.

Ideally, a concrete column should be designed such that the lateral load strength is controlled by flexure. However, even if most of the inelastic action is flexural, a column may not be sufficiently tough to sustain the imposed flexural deformations without failure.

Such failures are particularly common in older bridges. In the United States, the transverse reinforcement of reinforced concrete columns designed before 1971 commonly consists of #4 hoops (f = 13 mm) or ties at 12-in. (305-mm) spacing. Moreover, the ends of the transverse reinforcement rarely are anchored into the

Figure: *core of these columns. This amount and type of reinforcement provides negligible confinement to the concrete, particularly in large columns.*

Other detailing practices (in addition to providing little confinement) may lead to flexural failure in reinforced concrete columns. A common practice in Japan has been to terminate some of the longitudinal reinforcement within the column height. The resulting development length of the terminated reinforcement can be inadequate, and may lead to splitting failure along the terminated bars or to flexural and shear distress near the cutoff point.

Shear failures in reinforced concrete columns can be induced by interactions with "nonstructural" elements. These elements can decrease the distance between locations of flexural yielding, and therefore increase the shear demand for a column.

Lap splices of longitudinal reinforcement in older reinforced-concrete bridges may be vulnerable because, typically, the splices are short (on the order of 20 to 30 bar diameters), poorly confined, and are located in regions of high flexural demand. In particular, for construction convenience, splices are often located directly above a footing. With these details, the splices may be unable to develop the flexural capacity of the column, and they may be more vulnerable to shear failure. Despite these vulnerabilities, there is little field evidence of lap splice failures at the bases of bridge columns.

However, failures associated with welded splices and terminated longitudinal reinforcement were identified in the 1995 Hyogo-Ken Nanbu earthquake, as discussed previously. Concrete columns also can fail if the anchorage of the longitudinal reinforcement is inadequate. Such failures can occur both at the top of a column at the connection with the bent cap and at the bottom of a column at the connection with the foundation.

The column had been supported on a single 6-ft (1.8-m)-diameter, cast-in-drilled-hole pile. Other columns having hooked longitudinal reinforcement anchored in footings also failed with similar results in that earthquake. The consequences of foundation anchorage failures perhaps are larger in single-column bents than multicolumn bents, because the lateral-force resistance of single-column bents depends on the column developing its flexural strength at the base.

The record of steel column failures is sparse, because few bridges with steel columns have been subjected to strong earthquakes elsewhere than Japan. In the 1995 Hyogo-Ken Nanbu earthquake, failures apparently were associated with local buckling and subsequent splitting at welds or tearing of steel near the buckle. In columns with circular cross sections, the local buckling sometimes occurred at locations where section thicknesses changed. In rectangular columns, local buckling of web and flange plates was insufficiently restrained by small web stiffeners. A nearby column sustained local buckling at the base and tearing of the vertical welded seam between the two steel plates forming a corner of the column, suggesting the nature of failure that resulted in the collapse.

Beams

Beams traditionally have received much less attention than columns in seismic design and evaluation. In many bridges, the transverse beams are stronger than the columns because of gravity load requirements and composite action with the superstructure. Also, in many bridges, the consequences of beam failures are less severe than the consequences of column failures. In bridges with outriggers, however, the beams can be critical components of the bent and can be subjected to loadings that may result in failure. An example illustrating possible damage to an outrigger beam. This outrigger beam was monolithically framed with the superstructure and supporting column such that, under longitudinal load, significant torsion was required to be resisted by the outrigger portion of the beam. In some modern designs, torsion is reduced by providing nominal "pinned" connections between the beams and columns.

Joints

As with beams, joints traditionally have received little attention in seismic design, and they similarly may be exposed to critically damaging actions when the joints lie outside of the superstructure. Although joint failures occurred in previous earthquakes (e.g., Jennings), significant attention was not paid to joints until several spectacular failures were observed following the 1989 Loma Prieta earthquake. The occurrence of damage at the relatively large epicentral distance of approximately 60 miles (100 km) is attributed in part to site amplification and focusing of seismic waves as well as the vulnerability of the framing.

The collapse of the Cypress Street Viaduct during the Loma Prieta earthquake had more severe consequences. Failure of a concrete pedestal located just above the first-level jointled to the collapse of the upper deck on the lower deck, at a cost of 42 lives. Such pedestals are not common, but this collapse demonstrates that each earthquake has the potential to reveal a mode of failure that has not yet been considered routinely.

The Loma Prieta earthquake also identified an apparent weakness of a modern design. For example, damage occurred to the outrigger knee joints of the Route 980/880 connector, which had been constructed just a few years before the earthquake. This damage identified the need for special details in bridge construction, which has been the subject of important studies identified elsewhere in this book.

Abutments

The types of failures that can occur at abutments vary from one bridge to the next. The foundation type varies greatly (e.g., spread footing, pile-supported footing, drilled shafts), and the properties of the soil can be important, particularly if the soil liquefies during an earthquake. The situation is further complicated by the interaction of the backwalls, wingwalls, footings and piles with the surrounding soil. A common practice has been to treat abutments or abutment components as sacrificial elements, acting as fuses to relieve large seismic forces arriving at the stiff abutment. The occurrence of widespread and extensive damage in the 1994 Northridge earthquake suggests that an alternative approach might be economical.

In most seat-type abutments, longitudinal motion is unrestrained, because there is a joint at the interface of the superstructure and abutment backwall. This configuration is attractive, because it reduces the superstructure forces induced by temperature and shrinkage-induced displacements. The most important vulnerability of such abutments is span unseating, which can occur when there are large relative displacements between the superstructure and abutment seat. Abutment unseating failures are often attributable to displacement or rotation of the abutment, usually the result of liquefaction or lateral spreading.

Shear keys can be damaged also. Shear keys are components that restrain relative displacements (usually in the transverse direction) between the superstructure and the abutments. External shear keys are located outside of the superstructure cross section, while internal shear keys are located within the superstructure cross section. Since these elements are stocky, it is nearly impossible to make them ductile, and they will fail if their strength is exceeded.

Shear key failures were widespread during the 1994 Northridge earthquake. It appears that these failures can occur with small transverse displacement and little energy dissipation. Damage to internal shear keys usually is accompanied by damage to the interlocking backwall. In seat-type abutments, damage has also occurred in seat abutments due to pounding of backwalls by the superstructure.

In monolithic abutments, the superstructure is cast monolithically with the abutments. This configuration is attractive, because it reduces the likelihood of span unseating. However, the abutment can be damaged as the superstructure displaces in the longitudinal direction away from the abutment. Also, depending on the geometry and details of the abutment, the wingwall may serve as an external shear key. In such cases, the wingwall can fail in the same manner as an external shear key

Foundations

Reports of foundation failures during earthquakes are relatively rare, with the notable exception of situations in which liquefaction occurred. It is not clear whether failures are indeed that rare or whether many foundation failures are undetected because they remain underground. There are many reasons why older foundations might be vulnerable. Piles might have little confinement reinforcement, yet be subjected to large deformation demands. Older spread and pile-supported footings rarely have top flexural reinforcement or any shear reinforcement.

The 1995 Hyogo-Ken Nanbu earthquake resulted in extensive damage to superstructures and substructures above the ground, as reported elsewhere. The occurrence of that damage provided impetus to conduct extensive investigations of the conditions of foundation components. Along the older inland Route 3, an investigation of 109 foundations identified only cases of "small" flexural cracks in piles. Along the newer coastal Route 5, more extensive liquefaction occurred, resulting in lateral spreading in several cases. An investigation of 153 foundations for this route found cases of flexural cracks in piles where large residual displacements occurred, but the investigators found no spalling or reinforcement buckling. The absence of extensive damage was attributed to the spread of deformations along a significant length of the piles.

Foundation damage associated with liquefaction-induced lateral spreading has probably been the single greatest cause of extreme distress and collapse of bridges. The problem is especially critical for bridges with simple spans. The 1991 Costa Rica earthquake provides many examples of foundation damage.

Approaches

Even if the bridge structure remains intact, a bridge may bo placed out of service if the roadway leading to it settles significantly. For example, during the 1971 San Fernando earthquake and the 1985 Chile earthquake, settlement of the backfill abutments led to abrupt differential settlements in many locations. Such settlements can be large enough to pose a hazard to the travelling public. Approach or settlement slabs can be effective means of spanning across backfills.

Evaluation of Bridge Damage Data from Recent Earthquakes

Data on bridge damage from earthquakes is becoming increasingly more available. Such data, however, have not been systematically studied to evaluate damage characteristics and correlate these to observed or estimated local ground motions. In this task, data on bridge damage from the Loma Prieta and Northridge earthquakes were studied to correlate observed bridge damage to:

- Structural characteristics of a bridge
- Local ground motions
- Repair cost

In order to achieve these objectives, statistics on structural characteristics of bridges, ground shaking levels at bridge sites, damage characteristics and repair cost were obtained. Next, empirical damage probability matrices and fragility curves were developed from data on bridge damage. In addition, correlation between structural characteristics and observed damage were determined.

Despite the high ground motion levels observed in the 1994 Northridge, California earthquake, only about 3% of all the bridges in the area experienced major damage. The analyses of bridge damage data showed that concrete structures designed/built with older design standards were more prone to damage under seismic loading. The ground shaking level, skew angle, abutment type, pier type and span continuity showed the highest correlation with observed damage. The total repair cost for the damaged bridges was about $150,000,000, with repair and/or reconstruction of the collapsed structures forming a large portion of the total.

Bridge Damage Data Analysis Method

Characteristics of the Database: The database compiled for the Loma Prieta and Northridge earthquakes consists of four main types of data:

(a) structural characteristics,

(b) bridge damage,

(c) repair cost, and

(d) ground motion levels and soil characteristics at bridge sites.

A relational database management system (RDBMS), dBase™ , was used to compile and perform queries for data on bridge damage and structural characteristics of bridges. In addition, a geographic information system (GIS), Arc/Info™ , was used to obtain the ground motion levels at each bridge site.

Structural Characteristics: Structural characteristics compiled for the groups of bridges that were exposed to ground shaking included abutment type, number of spans, type of superstructure and substructure, length and width of the bridge, skew, number of hinges at joints and bents, abutment and column foundation types, retrofit history, and design year to represent design standards, such as column reinforcement and seat width. These structural characteristics were obtained from the Structural Maintenance System (SMS) database compiled and managed by Caltrans (Caltrans, 1993). In addition, more detailed information was obtained from Caltrans for some of the damaged bridges. Caltrans is currently in the process of compiling a database that includes information on abutment, bent/pier/column and bent/pier/footing details, such as seat width and type of bearings, footing type and column/footing connection. However, only about 15% of all the California bridges are currently in this database, thus this detailed information could not be used in the statistical analyses.

Bridge Damage: Detailed damage descriptions and the corresponding damage states were compiled for bridges damaged in the two earthquakes. The damage descriptions were obtained mainly from bridge damage reports compiled by Caltrans (1989, 1994). For the Northridge earthquake, these descriptions were cross-referenced with those provided by Buckle (1994), EERI (1995), and Yashinsky (1995). Judgement was used to treat inconsistencies in the interpretation of the observed damage data.

The database on bridge damage specifies two damage states (*minor, major*) for bridges damaged in the Loma Prieta earthquake, and four damage states (*minor, moderate, major* and *collapse*) for those damaged in the Northridge earthquake. The bridge damage data were used in correlation studies to obtain ground motion-damage relationships.

Damage State Definitions: Currently, no guidelines for evaluating physical bridge damage exist. The terms *minor*, *moderate* and *major* damage were subjective. Definitions of damage states for columns, abutments, and joints and connections for concrete bridges are proposed which were developed based on the observed bridge damage in the Northridge earthquake (Basöz, 1996). A questionnaire was prepared to acquire expert opinion on the proposed damage state definitions and given to bridge engineers at Caltrans. The feedback provided by the bridge engineers will be used to modify some of the damage state definitions.

Repair Cost: The estimated repair cost values for damaged bridges were obtained from supplementary bridge reports compiled by Caltrans following each earthquake. The database includes total estimated repair cost and more detailed information on repair work and cost for each bridge that was repaired.

The repair cost ratio, defined as the ratio of repair cost to replacement cost of a bridge, was calculated for all the damaged bridges. The replacement cost of a bridge was estimated to be \$90/ft^2 based on the 1995 cost books.

Ground Motion Levels: In addition to structural characteristics, soil type at each bridge site and peak ground acceleration (PGA) levels observed in the two earthquakes were compiled. In order to obtain empirical ground motion-damage relationships for the set of bridges damaged in the Northridge earthquake, two sets of peak ground acceleration (PGA) values were used as the ground motion levels:

(a) PGA values reported by USGS (1994), which were obtained from the contours of observed PGA recordings in the horizontal direction and

(b) PGA values reported by WCFS (1995) that were obtained from the contours of the average of the PGA values measured in the E-W and N-S directions.

Recorded ground motion levels were used to scale the parameters of empirical Greenís functions which were used in simulating the ground shaking levels (Somerville et al., 1996). The PGA value at a given bridge site was obtained within GIS by overlaying the ground shaking map and the bridge location map.

Subsequently, the highest PGA values obtained at a bridge site were 1.55g and 0.66g for the USGS and WCFS maps, respectively. Since the PGA levels from the two data sets varied considerably, correlation studies were performed for both data sets. Because no contour maps were available for the PGA levels observed during the Loma Prieta earthquake, attenuation relationships were used to estimate the level of ground shaking at each bridge site.

Classification of Bridges

The compiled inventory of bridges was reviewed to: (a) select bridges to be used in correlation studies, (b) select structural characteristics (attributes) that best describe the seismic response of bridges, and (c) verify the correctness of the attribute values included in the bridge inventory database.

Data Sets: Several data sets were used for statistical analyses. All the analyses were performed for the state bridges since most of the reported damage in both earthquakes pertain to state bridges. First, all highway state bridges were selected and gathered in the *highway bridge data set*, and statistics on design year and ground shaking levels were obtained. Most of the bridge damage pertained to concrete structures. The number of damaged steel bridges was not large enough for statistical analysis. Therefore, concrete bridges were selected from the *highway bridge data set*.

One of the objectives of this research was to identify the effect of various structural characteristics on bridge damage. In order to study the effect of structural component types on bridge damage, such as effect of abutment type (monolithic or non-monolithic), and effect of number of columns per bent, bridges with single abutment type and one column bent type were compiled in a database called the *homogeneous data set*.

That is, in order to determine the effect of each characteristic, only bridges with homogeneous structural characteristics were selected from the *concrete highway bridge data set*. For example, a bridge with a seat type abutment (non-monolithic), and a diaphragm type (monolithic), was defined as a *heterogeneous* bridge and excluded from the *homogeneous data set*. Similarly, a bridge with both multiple and single column bents was defined as a heterogeneous bridge, and was excluded from the *homogeneous data set*. Bridges with incomplete information were also excluded.

Another criterion in the data selection relates to the correlation analyses. A complete data set for correlation analyses requires that all the bridges exposed to a given ground shaking level be included. In order to satisfy this requirement, a minimum PGA level was selected as a threshold value. This PGA level was determined based on the available ground motion maps. The data set that satisfies this condition was extracted from the *homogeneous data set* and is referred to as the *correlation data set*.

The bridges in the *correlation data set* were grouped first by the superstructure type and substructure material. Then, these bridges were further classified into subcategories based on other structural characteristics, such as number of spans, abutment type, column bent type and span continuity.

The classification scheme used in this task was adapted from the bridge classification developed by Basöz and Kiremidjian (1996) under another NCEER project. The damaged bridges were classified to group bridges together that were expected to experience similar damage levels under a given seismic loading. The correlation studies were also carried out using the bridge classification defined by the National Institute of Building Sciences (NIBS) Manual (RMS, 1995).

Reliability of the Database: Caltrans currently has two database systems: the first one (SMS) follows the Federal Highway Administration (FHWA) National Bridge Inventory System but is more detailed, and the second one (BIRIS) is a database that stores bridge books and drawings for all the bridges. The bridge books include detailed reports from each bridge inspection. For some of the 25,000 bridges in the state of California, discrepancies exist between the two databases.

The two databases were compared to verify the correctness of the attribute values for some of the bridges that were damaged in the Northridge earthquake. The abutment type and column bent type were found to be more likely to have errors than the other attributes of interest for this research.

For example, in only a few cases (for 2 to 3% of the damaged bridge data set), the values of the design year and skew attributes were found to be incorrect. Where there were discrepancies between the two databases, the structural plans were investigated with the assistance of bridge engineers from Caltrans to determine the correct attribute values. In order to evaluate the effect of error inherent in the database on the results of the correlation studies, several analyses were performed using sample data sets with corrected and uncorrected data.

Correlation Studies

Correlation analyses were performed using the data on bridge damage and repair cost. The correlation studies were performed for the following objectives:

- To determine the structural characteristics that best represent damage such that bridges can be grouped using these characteristics,
- To obtain ground motion-damage relationships for bridges with similar structural characteristics,
- To obtain ground motion-repair cost ratio relationships to estimate direct economic losses due to damage to bridges,
- To correlate damage and repair cost ratio.

The data on bridge damage were compiled in the form of damage matrices, i.e., the number of bridges with each level of observed damage at different PGA levels.

Then, the damage probability matrices (DPMs), i.e., the probability of being in a damage state given the ground motion level were obtained for each group of bridges.

The damage matrices were used as input data to logistic regression analysis to obtain empirical fragility curves both unconditional and conditional on damage.

Similar procedures were used to obtain empirical fragility curves for the repair cost ratio.

Comparison of observed damage data to currently available ground motion-damage relationships (ATC, 1985; RMS, 1995) were presented in Basöz et al. (1997).

Example Statistics and Results

Some of the results based on data from the Northridge earthquake are presented in this section.

The bridges in the Greater Los Angeles area including Los Angeles, Ventura, Riverside, and Orange Counties, were exposed to ground shaking during the 1994 Northridge earthquake.

The number of state and local bridges and the number of damaged state bridges in each of the four counties. A database that includes state and local bridges for the four counties was extracted from the Bridge Maintenance Database compiled by Caltrans (1993).

Structural characteristics such as structural type and material, number of spans, abutment type, span continuity, design year indicating the seat width and column longitudinal reinforcement, substructure type, skew, and foundation type were included in this database.

Table: Distribution of State and Local Bridges and the Number of Damaged State Bridges in the Greater Los Angeles Area				
County	***No. of State Bridges***	***No. of Local Bridges***	***Total No. of Bridges***	***No. of Damaged Bridges***
Los Angeles	2,097	1,553	3,650	228
Riverside	644	338	982	-
Orange	463	505	968	-
Ventura	329	175	504	5
Total	3,533	2,571	6,104	233

Bridge damage from the Northridge earthquake pertained mostly to state bridges in Los Angeles and Ventura Counties. Bridges in these two counties also experienced much higher accelerations than those in Riverside and Orange Counties. A total of 63 bridges were exposed to peak ground acceleration (PGA) levels of 0.15g or higher in Riverside and Orange Counties. 3,533 state and 2,571 local bridges are located in the four counties. Of the 3,533 state bridges, 3,318 (1,902 bridges in Los Angeles County, 312 in Ventura County, 462 in Orange County and 642 in Riverside County) carry highway traffic and were included in the highway bridge data set. The number of bridges in the highway bridge data set by superstructure type and substructure material.

Table: Distribution of Highway Bridges in Los Angeles, Ventura, Orange and Riverside Counties by Structural and Material Type				
	Concrete	***Steel***	***N/A***	***Timber***
Concrete Girder	2,396	9[1]	708	0
Steel Girder	119	3	32	0
Truss	4	0	0	0
Tunnel	0	0	4	0
Timber	0	1	0	4
Arch	12	0	15	0
Suspension	1	0	0	0
Unknown	4	0	6	0
[1] *These nine bridges have concrete slab type superstructure and have both concrete and steel column bents.*				

1- *These nine bridges have concrete slab type superstructure and have both concrete and steel column bents.*

The distribution of these bridges by design year. Seventy seven percent of the bridges in the highway bridge data set were designed by pre-1971 design standards. The majority of bridges in the four counties were concrete structures as were more than 85% of the damaged bridges.

Therefore, the statistical analyses were conducted mainly for concrete bridges. The recorded PGA values (USGS, 1994) and the bridges in Los Angeles, Ventura, Orange and Riverside counties. The PGA value at a given bridge site was obtained within GIS.

Table: *Distribution of Estimated Repair Cost by Damage State*

Damage State	*Number of Bridges*	*Estimated Repair Cost*
Collapse	6	$121,765,750
Major	47	$18,398,057
Moderate	96	$6,895,731
Minor	85	$446,950

For the repair cost data, a database was compiled from the supplementary bridge damage reports provided by Caltrans. A total of about $150,000,000 was reported as repair cost in these reports. The total repair cost for the six collapsed bridges constitute 75% of the repair cost of all damaged bridges. The database includes total estimated repair cost and more detailed information on repair work and cost for 130 bridges in Los Angeles and Ventura counties. The estimated repair cost of damaged bridges by damage state. Fragility curves for repair cost ratio for multiple span bridges.

Concluding Remarks

The results from this task can be used to assist in making decisions about mitigation, such as prioritizating bridges for seismic retrofitting, and for post-earthquake response and recovery activities. More specifically, the areas that can benefit from the results of this task include the following:

Database of Bridge Damage and Structural Characteristics: A comprehensive database will be available for the most recent two major earthquakes in the United States. This database includes information on bridge damage and the structural characteristics that are important for vulnerability assessment of bridges. This type of a database provides an essential base to improve our understanding of bridge damage from past earthquakes.

Classification of Bridges and Ground Motion-damage Relationships: Currently available bridge classes and the corresponding ground motion-damage relationships are rudimentary and do not properly estimate the observed damage from the Northridge earthquake. The method used in this research utilizes the observed damage data to develop empirical fragility curves which can and should be improved as more data become available. PGA levels, skew, span continuity, abutment type, and number of spans correlated well with observed damage. These are the characteristics used in the classification by Basöz and Kiremidjian (1996) and suggests that data shows good agreement with the structural characteristics used in that bridge classification.

Damage State Definitions: Post-earthquake damage assessment is an important area that needs to be addressed for efficient and effective emergency response management. The damage states proposed in this research can be used to develop a post-earthquake investigation form that will assist in compiling bridge damage.

2

Dynamic Analysis

Dynamic Analysis of Base-Isolated Highway Bridge

Four-Span Isolated Bridge

This example uses ALADDIN for a three dimensional nonlinear time-history analysis of a 4 span base isolated highway bridge. The bridge model is taken from the text of Priestley et al.

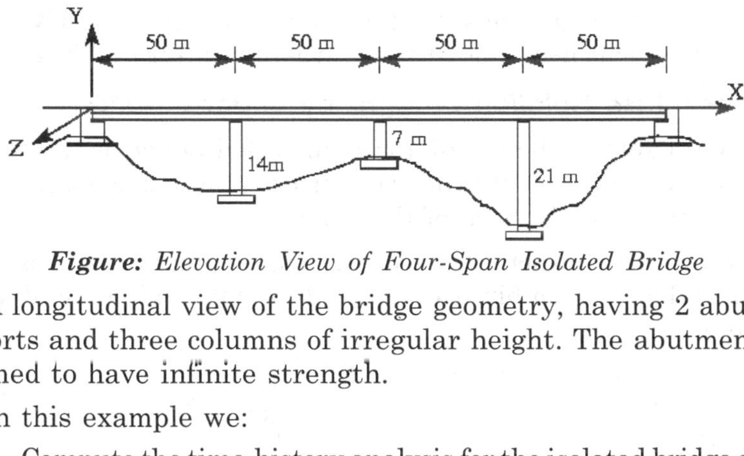

Figure: *Elevation View of Four-Span Isolated Bridge*

A longitudinal view of the bridge geometry, having 2 abutment supports and three columns of irregular height. The abutments are assumed to have infinite strength.

In this example we:

1. Compute the time-history analysis for the isolated bridge subject to a 10 second segment of the Elcentro ground motion scaled to a peak ground acceleration 0.5g.
2. Recompute the time-history analysis for a non-isolated bridge subject to a 10 second segment of the Elcentro ground motion scaled to a peak ground acceleration 0.5g.
3. Compute the natural periods of vibration for the isolated and non-isolated highway bridges.

4. Estimate the ductility demand on the individual isolators as well as the overall ductility of the highway bridge structure.
5. Conduct and energy-balance calculation for the flow of energy through the bridge during the time-history response.

Section and Material Properties

All of the piers have identical cross sectional properties, but the reinforcement is detailed so that the piers have approximately the same lateral force resistance.

Table: *Pier and Isolator Capacities*

	Secant Stiffness (kN/m)	***Yield Strength (kN)***
Isolators 1 and 5	5300	2100
Isolator 2	16100	2100
Isolator 3	11500	2100
Isolator 4	47000	2100
Pier 2	30700	
Pier 3	124400	
Pier 4	13650	

The isolators are assumed to have elastic-perfectly plastic behaviour, which means no strain hardening after yielding.

How Does Base Isolation of Highway Bridges Work?

When a seismic base isolator is installed between the top of a bridge column and beneath the deck structure, the parameters of the column combine with those of the isolator.

The overall characteristics of the superstructure and isolation system reflect a combination of individual composite dynamic parameters.

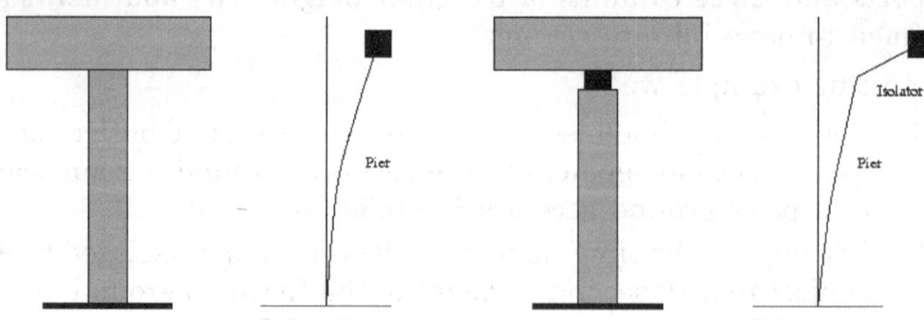

FIXED - BRIDGE ISOLATED - BRIDGE

Figure: *Mode Shapes in Lateral Direction*

For most practical situations, the superstructure mass is much greater than the column mass, and hence the first mode of vibration is dominated by displacements of the superstructure. The second mode, in comparison, involves lateral motion of the top of the column with little movement of the superstructure.

Finite Element Model

The finite element mesh has 22 nodes and 21 elements. The latter are a combination of FRAME_3D and FIBER_3D elements.

***Table:** Use of Finite Elements in Bridge Model*

Element Nos	*Description*	*Element Type*
1 to 4	Bridge Deck Attribute	FRAME_3D
5 and 9	Isolator Attribute 1	FIBER_3D
6	Isolator Attribute 2	FIBER_3D
7	Isolator Attribute 3	FIBER_3D
8	Isolator Attribute 4	FIBER_3D
10 to 13	Column 2	FRAME_3D
14 and 15	Column 3	FRAME_3D
16 to 21	Column 4	FRAME_3D

The deck and columns are modelled with elastic 3-D frame elements (i.e., element type FRAME_3D). Only the isolators are modelled as 3-D nonlinear fibre elements with bi-linear shear-displacement response (i.e., element type FIBER_3D).

The lower ends of the piers and abutments are fully fixed. We also assume that there are no longitudinal deformations and twisting at the both ends of the bridge deck. In total there are 21 elements, 8 different element attributes and 98 degrees of freedom in this model.

Dynamic Analysis

Earthquake Ground Motion

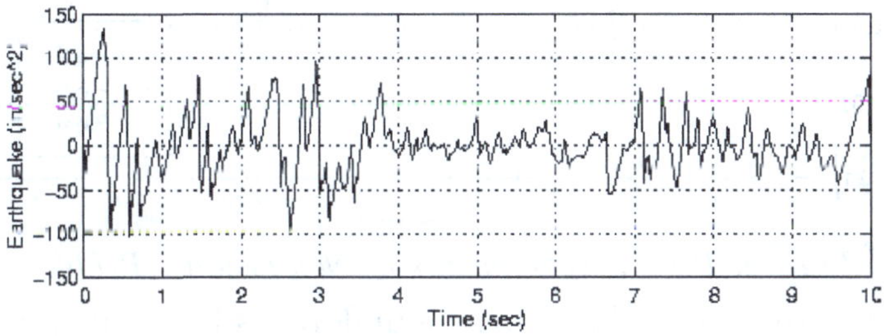

***Figure:** 1940 El Centro S00E Record*

We choose the 1940 El Centro S00E record as the input earthquake with the peak ground acceleration scaled to 0.5g.

The deck and piers are assumed to remain elastic during the dynamic responses.

The earthquake ground motion is defined by the block of code:

```
Elcentrol = ColumnUnits( [
     14.56;     13.77;      6.13;      3.73;      1.32;     -6.81;    -16.22;   -
22.41;
    -26.67;    -24.10;    -21.86;    -14.02;     -1.62;      1.67;     18.12;
39.26;
     .... details of ground acceleration record removed ....
     -3.05;     -6.01;     -6.08;     -6.40;     -6.70;     -5.70;     -6.14;   -
8.41;
    -10.05;    -12.35;    -15.72;      0.00
   ], [cm/sec/sec] );
```

Each number in the array Elcentrol corresponds to the digitized ground motion acceleration measured in cm/sec/sec and spaced at 0.02 seconds.

Displacements

The deck displacements versus time for the non-isolated and isolated bridge structures.

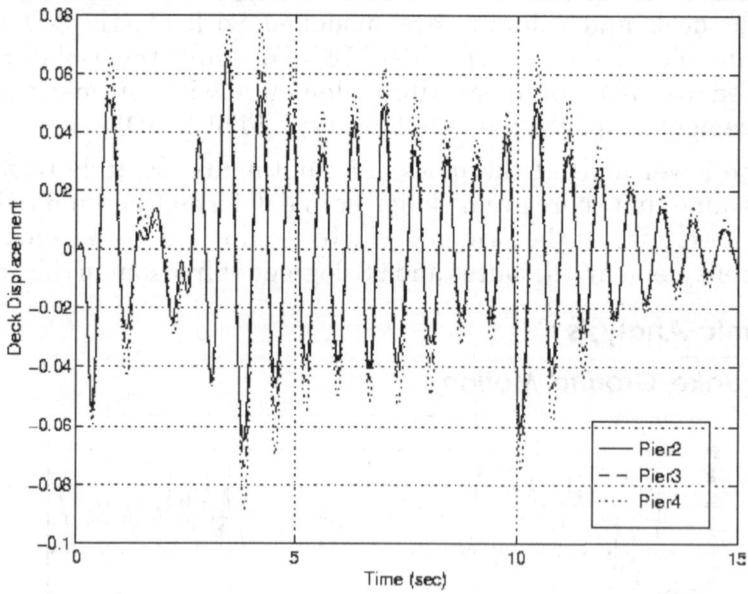

Figure: *Deck Displacements of Non-Isolated Bridge*

As expected, the isolated bridge undergoes larger displacements than the non-isolated bridge. For example, at Pier 4 the maximum displacement of the isolated bridge is about 3.2 times larger than the maximum displacement of the non-isolated bridge. Also notice that because the isolators undergo nonlinear deformations, the isolated bridge has a 0.02 m residual displacement after the earthquake response has ceased.

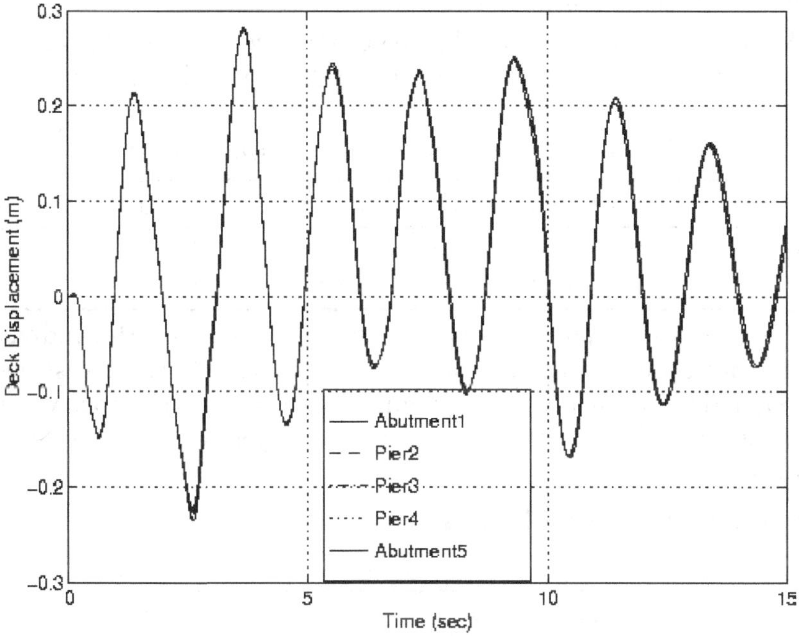

Figure: *Deck Displacements of Isolated Bridge*

Ductility Demand

Capacity design principles for base isolated bridges dictate that hysteretic energy dissipation should be confined to the isolation system. Only under extreme circumstances should the piers deform into the inelastic range.

The force-displacement response of a typical isolator, in this case, the isolator at Pier 4.

The isolator ductility "mu_d" may be defined as:

$$\text{mu\_d} = \frac{\text{delta\_p}}{\text{delta\_de}} + 1$$

Here "delta_de" and "delta_p" are the elastic and plastic deformation components of the isolator, respectively (hence, the total isolator displacement is "delta_de + delta_p").

The effective ductility for the combined pier-isolator system is:

$$mu\_g = \frac{delta\_p}{delta\_s + delta\_de} + 1$$

where "delta_de" and "delta_p" are as previously defined, and "delta_s" is the elastic deformation of the pier.

With this notion in place, Table summarizes the isolator and pier-isolator ductility demands occurring at the bridge abutments and pier supports.

***Table:** Ductility Demands on Pier-Isolator System*

Ductility	*Abutments 1, 5*	*Pier 2*	*Pier 3*	*Pier 4*
mu_d	1.41	1.64	1.45	2.78
mu_g	1.41	1.42	1.41	1.40

The important point to note is that by designing the individual isolators for different ductility demands, the overall bridge system can be designed so that the ductilty demand on the pier-isolator system is about the same. This design strategy helps to regularize the structural response.

Energy Balance Calculations

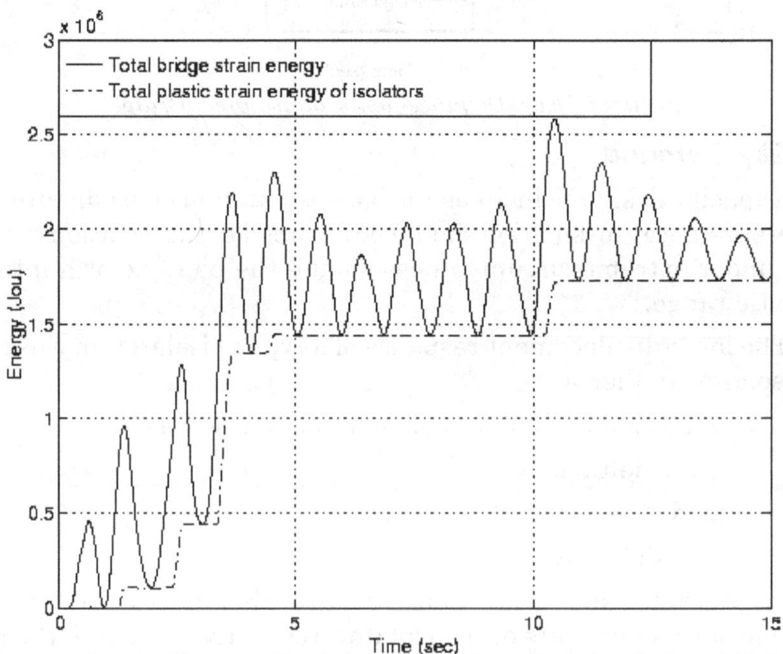

***Figure:** Strain Energy (Joules) versus Time (sec) for Isolated Bridge*

Components of strain energy (Joules) versus time (sec) for the isolated bridge. The dashed line shows the time-variation in plastic strain-energy (or hysteretic energy) dissipated by the isolators. The solid line shows the time-variation in total strain energy for the isolated bridge.

The difference between the solid and dashed lines corresponds to the strain energy stored in the bridge piers and deck.

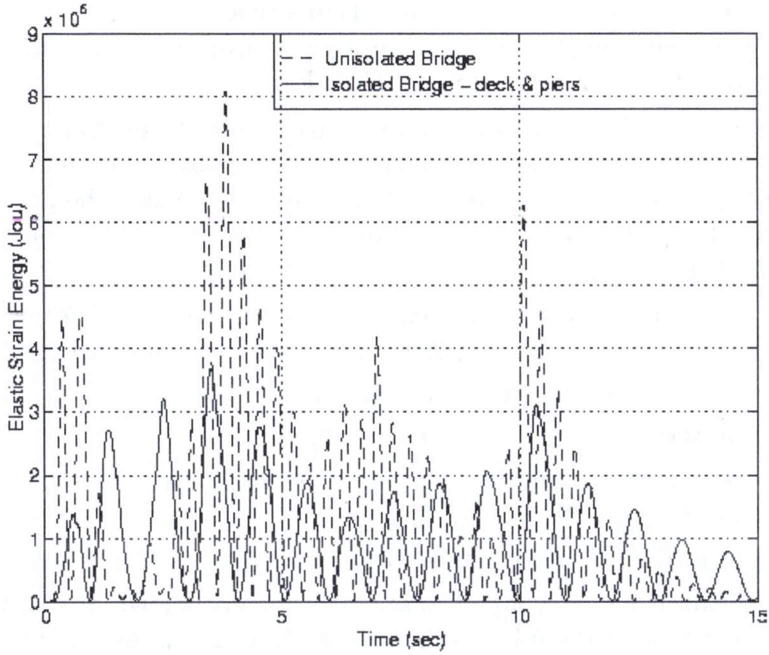

Figure: *Strain Energy of Deck and Piers*

Compares the strain energy in the deck and piers for the isolated and non-isolated structures. Because the strain energy components are roughly proportional to the square of the bridge displacements, higher strain levels of energy in the bridge components means higher possibility of damage. Clearly shows that the peak values of strain energy in the isolated bridge deck are lower than those in the conventionally supported bridge deck. Hence, we conclude that the isolation system protects the bridge superstructure, as required.

Nonlinear Analysis of Bridge Structures

In recent years, nonlinear bridge analysis has gained a greater momentum because of the need to assess inelastic structural behaviour under seismic loads. Common seismic design philosophies for ordinary bridges allow some degree of damage without collapse. To control and evaluate damage, a postelastic nonlinear analysis is required. A nonlinear analysis is complex and involves many simplifying assumptions. Engineers must be familiar with those complexities and assumptions to design bridges that are safe and economical.

Many factors contribute to the nonlinear behaviour of a bridge. These include factors such as material inelasticity, geometric or second-order effects, nonlinear soil–foundation–structure inter-action, gap opening and closing at hinges and abutment locations, time-dependent effects due to concrete creep and shrinkage, etc.

Because of the wide range of structural possibilities, this Spotter's Guide shows only the most common fixed (non-movable) bridge types. Other types are listed in the Bridge Terminology page. The drawings are not to scale. Additional related info is found on the other Terminology pages which are linked to the left.

The four main factors are used in describing a bridge. By combining these terms one may give a general description of most bridge types.

- span (simple, continuous, cantilever),
- material (stone, concrete, metal, etc.),
- placement of the travel surface in relation to the structure (deck, pony, through),
- form (beam, arch, truss, etc.).

The three basic types of spans are shown below. Any of these spans may be constructed using beams, girders or trusses. Arch bridges are either simple or continuous (hinged). A cantilever bridge may also include a suspended span.

Examples of the three common travel surface configurations are shown in the Truss type drawings below. In a Deck configuration, traffic travels on top of the main structure; in a Pony configuration, traffic travels between parallel superstructures which are not cross-braced at the top; in a Through configuration, traffic travels through the superstructure (usually a truss) which is cross-braced above and below the traffic.

Beam and Girder Types

Simple deck beam bridges are usually metal or reinforced concrete. Other beam and girder types are constructed of metal. The end section of the two deck configuration shows the cross-bracing commonly used between beams. The pony end section shows knee braces which prevent deflection where the girders and deck meet.

One method of increasing a girder's load capacity while minimizing its web depth is to add haunches at the supported ends. Usually the centre section is a standard shape with parallel flanges; curved or angled flanged ends are riveted or bolted using splice plates. Because of the restrictions incurred in transporting large beams to the construction site, shorter, more manageable lengths are often joined on-site using splice plates.

Many modern bridges use new designs developed using computer stress analysis. The rigid frame type has superstructure and substructure which are integrated. Commonly, the legs or the intersection of the leg and deck are a single piece which is riveted to other sections.

Orthotropic beams are modular shapes which resist stress in multiple directions at once. They vary in cross-section and may be open or closed shapes.

Arch Types

There are several ways to classify arch bridges. The placement of the deck in relation to the superstructure provides the descriptive terms used in all bridges: deck, pony, and through.

Also the type of connections used at the supports and the midpoint of the arch may be used - counting the number of hinges which allow the structure to respond to varying stresses and loads. A through arch is shown, but this applies to all type of arch bridges.

Another method of classification is found in the configuration of the arch. Examples of solid-ribbed, brace-ribbed (trussed arch) and spandrel-braced arches are shown. A solid-ribbed arch is commonly constructed using curved girder sections. A brace-ribbed arch has a curved through truss rising above the deck. A spandrel-braced arch or open spandrel deck arch carries the deck on top of the arch.

Some metal bridges which appear to be open spandrel deck arch are, in fact, cantilever; these rely on diagonal bracing. A true arch bridge relies on vertical members to transmit the load which is carried by the arch.

The tied arch (bowstring) type is commonly used for suspension bridges; the arch may be trussed or solid. The trusses which comprise the arch will vary in configuration, but commonly use Pratt or Warren webbing. While a typical arch bridge passes its load to bearings at its abutment; a tied arch resists spreading (drift) at its bearings by using the deck as a tie piece.

Masonry bridges, constructed in stone and concrete, may have open or closed spandrels A closed spandrel is usually filled with rubble and faced with dressed stone or concrete. Occasionally, reinforced concrete is used in building pony arch types.

Truss - simple Types

A truss is a structure made of many smaller parts. Once constructed of wooden timbers, and later including iron tension members, most truss bridges are built of metal. Types of truss bridges are also identified by the terms deck, pony and through which describe the placement of the travel surface in relation to the superstructure. The king post truss is the simplest type; the queen post truss adds a horizontal top chord to achieve a longer span, but the centre panel tends to be less rigid due to its lack of diagonal bracing.

Covered Bridge Types (Truss)

Covered bridges are typically wooden truss structures. The enclosing roof protected the timbers from weathering and extended the life of the bridge.

One of the more common methods used for achieving longer spans was the multiple kingpost truss. A simple, wooden, kingpost truss forms the centre and panels are added symmetrically. With the use of iron in bridge construction, the Howe truss - - in its simplest form - - appears to be a type of multiple kingpost truss.

Stephen H. Long (1784-1864) was one of the U.S. Army Topographical Engineers sent to explore and map the United States as it expanded westward. While working for the Baltimore and Ohio Railroad, he developed the X truss in 1830 with further improvements patented in 1835 and 1837. The wooden truss was also known as the Long truss and he is cited as the first American to use mathematical calculations in truss design.

Theodore Burr built a bridge spanning the Hudson River at Waterford, NY in 1804. By adding a arch segments to a multiple kingpost truss, the Burr arch truss was able to attain longer spans. His truss design, patented in 1817, is not a true arch as it relies on the interaction of the arch segments with the truss members to carry the load. There were many of this type in the Pittsburgh area and they continue to be one of the most common type of covered bridges. Many later covered bridge truss types used an added arch based on the success of the Burr truss.

The Town lattice truss was patented in 1820 by Ithiel Town. The lattice is constructed of planks rather than the heavy timbers required in kingpost and queenpost designs. It was easy to construct, if tedious. Reportedly, Mr. Town licensed his design at one dollar per foot - - or two dollars per foot for those found not under license. The second Ft. Wayne railroad bridge over the Allegheny River was an unusual instance of a Town lattice constructed in iron.

Herman Haupt designed and patented his truss configuration in 1839. He was in engineering management for several railroads including the Pennsylvania Railroad (1848) and drafted as superintendent of military railroads for the Union Army during the Civil War. The Haupt truss concentrates much of its compressive forces through the end panels and onto the abutments.

Other bridge designers were busy in the Midwest. An OhioDOT web page cites examples of designs used for some covered bridges in that state. Robert W. Smith of Tipp City, OH, received patents in 1867 and 1869 for his designs. Three variations of the Smith truss are still standing in Ohio covered bridges.

Reuben L. Partridge received a patent for his truss design which is appears to be a modification of the Smith truss. Four of the five Partridge truss bridges near his home in Marysville, Union County, OH, are still in use.

Horace Childs' design of 1846 was a multiple king post with the addition of iron rods. The Childs truss was used exclusively by Ohio bridge builder Everett Sherman after 1883.

Truss - Pratt Variations

The Pratt truss is a very common type, but has many variations. Originally designed by Thomas and Caleb Pratt in 1844, the Pratt truss successfully made the transition from wood designs to metal. The basic identifying features are the diagonal web members which form a V-shape. The centre section commonly has crossing diagonal members.

Additional counter braces may be used and can make identification more difficult, however the Pratt and its variations are the most common type of all trusses.

Charles H. Parker modified the Pratt truss to create a "camelback" truss having a top chord which does not stay parallel with the bottom chord. This creates a lighter structure without losing strength; there is less dead load at the ends and more strength concentrated in the centre. It is somewhat more complicated to build since the web members vary in length from one panel to the next.

When additional smaller members are added to a Pratt truss, the various subdivided types have been given names from the railroad companies which most commonly used each type, although both were developed by engineers of the Pennsylvania Railroad in the 1870s.

The Whipple truss was developed by Squire Whipple as stronger version of the Pratt truss. Patented in 1847, it was also known as the "Double-intersection Pratt" because the diagonal tension members cross two panels, while those on the Pratt cross one. The Indiana Historical Bureau notes one bridge as being a "Triple Whipple" — possibly the only one — built with the thought that if two are better than one, three must be stronger yet.

The Whipple truss was most commonly used in the trapezoidal form — straight top and bottom chords — although bowstring Whipple trusses were also built.

The Whipple truss gained immediate popularity with the railroads as it was stronger and more rigid than the Pratt. It was less common for highway use, but a few wrought iron examples survive. They were usually built where the span required was longer than was practical with a Pratt truss.

Further developments of the subdivided variations of the Pratt, including the Pennsylvania and Baltimore trusses, led to the decline of the Whipple truss.

Truss - Warren Variations

A Warren truss, patented by James Warren and Willoughby Monzoni of Great Britain in 1848, can be identified by the presence of many equilateral or isoceles triangles formed by the web members which connect the top and bottom chords. These triangles may also be further subdivided. Warren truss may also be found in covered bridge designs.

Truss - other Types

The other truss types shown are less common on modern bridges. A Howe truss at first appears similar to a Pratt truss, but the Howe diagonal web members are inclined toward the centre of the span to form A-shapes. The vertical members are in tension while the diagonal members are in compression, exactly opposite the structure of a Pratt truss. Patented in 1840 by William Howe, this design was common on early railroads.

The three drawings show various levels of detail. The thicker lines represent wood braces; the thinner lines are iron tension rods. The Howe truss was patented as an improvement to the Long truss which is discussed with covered bridge types.

Friedrich August von Pauli (1802-1883) published details of his truss design in 1865. Probably the most famous Pauli truss, better known as the lenticular truss — named because of the lens shape, is Pittsburgh's Smithfield Street Bridge. Its opposing arches combine the benefits of a suspension bridge with those of an arch bridge. But like the willow tree, some of its strength is expressed in its flexibility which is often noticeable to bridge traffic.

Before the use of computers, the interaction of forces on spans which crossed multiple supports was difficult to calculate. One solution to the problem was developed by E. M. Wichert of Pittsburgh, PA, in 1930. By introducing a open, hinged quadrilateral over the intermediate piers, each span could be calculated independently. The first Wichert truss was the Homestead High Level Bridge over the Monongahela River in 1937.

The composite cast and wrought iron Bollman truss was common on the Baltimore and Ohio Railroad. Of the hundred or so following Wendell Bollman's design, the 1869 bridge at Savage, MD, is perhaps the only intact survivor. Some of the counter bracing inside the panels has been omitted from the drawing for clarity.

Also somewhat common on early railroads, particularly the B&O, was the Fink truss — designed by Albert Fink of Germany in the 1860s.

Cantilever Types – Truss

A cantilever is a structural member which projects beyond its support and is supported at only one end. Cantilever bridges are constructed using trusses, beams, or girders. Employing the cantilever principles allows structures to achieve spans longer than simple spans of the same superstructure type. They may also include a suspended span which hangs between the ends of opposing cantilever arms.

Some bridges which appear to be arch type are, in fact, cantilever truss. These may be identified by the diagonal braces which are used in the open spandrel. A true arch bridge relies on vertical members to transfer the load to the arch. Pratt and Warren bracing are among the most commonly used truss types.

The classic cantilever design is the through truss which extends above the deck. Some have trusses which extend both above and below the deck. The truss configuration will vary.

Suspension Types

The longest bridges in the world are suspension bridges or their cousins, the cable-stayed bridge. The deck is hung from suspenders of wire rope, eyebars or other materials. Materials for the other parts also vary: piers may be steel or masonry; the deck may be made of girders or trussed. A tied arch resists spreading (drift) at its bearings by using the deck as a tie piece.

Though Pittsburgh has been a pioneer in bridge design and fabrication, it has had few suspension bridges. The Pennsylvania Mainline Canal entered the city on John Roebling's first wire-rope suspension bridge in 1845 (replacing a failing 1829 wooden structure). A similar structure still stands at Minnisink Ford, NY, crossing the Delaware River. Roebling and his son Washington Roebling, later famous in building the Brooklyn Bridge, began their work in Saxonburg, PA, north of Pittsburgh.

Studies related to the design and analysis of curved and skewed steel bridges have focused on modelling and design for static and pseudo-static loads, and only a few investigations have looked at seismic behaviour. To design and assess curved steel bridges in high and moderate seismic zones, it is of interest to more extensively examine seismic analysis methods so that reliable 3D modelling approaches are developed. Studies have been undertaken that applied modelling approaches to predict the seismic response of straight steel girder bridges. Similar simplified modelling approaches have been proposed for curved steel bridges, but the approaches were applied to static events. These studies have shown that modelling using a 3D approach can provide improved accuracy relative to line girder analyses by incorporating member depths. For a curved and skewed bridge, where significant lateral displacements may be induced at the bearings under a seismic event, modelling structural component depths would be assumed to be important.

For these reasons, a 3D modelling approach is used herein to investigate seismic responses of a curved steel I-girder bridge system with skewed supports. The 3D approach is applied to a three-span continuous curved steel I-girder bridge system in the United States. Following the approach recommended by previous research, the bridge is modelled using elastic frame elements for the I-girders, truss elements for the cross-frames, and elastic frame elements for the deck. Preliminary seismic responses at the bearings are presented for the bridge under El Centro ground motions.

Modelling Approach

All elements used for 3D model were generalized using Open Sees. Curved bridge framing was represented using frame elements with lumped masses being placed at each node, with those masses calculated using tributary dimensions. Model construction initiated with calculation of superstructure and substructure section properties. Superstructure included girder, cross frame, concrete deck, and rigid link element, while substructure included pier column, cap, abutment, and footing. Spherical bearings were modelled in Open Sees using Zero Length elements.

All rotational degrees of freedom for the bearings, which accommodated rotations about various axes, were unrestrained. To simulate the bearings' moment-rotational behaviour, a combination of different material models available in Open Sees was utilized. Appropriate nominal material properties were used for the steel and concrete.

Seismic Response of Curved and Skewed Bridge

Preliminary results from seismic analyses of the curved and skewed bridge using the 3D model are presented. These results focus on seismic bearing response as a result of longitudinal and transverse earthquake loadings.

Curved and Skewed Longitudinal Earthquake Loadings

Since the structure being examined is horizontally curved and rests on skewed supports, directions both parallel to and perpendicular to the skewed supports were identified as those for the ground motions. Applying ground motions in this fashion has been shown to be preferred for inducing critical skewed bridge response. To apply these motions, two "longitudinal directions" were defined. The direction tangential to the chord of each curved girder at the abutment and/or pier was referred to as the "curved longitudinal" direction. To capture critical superstructure response, El Centro ground motions, which had a peak horizontal ground acceleration of 0.313 g, were applied to the bridge in the curved longitudinal direction initially and then the skewed longitudinal direction. Bearing rotations were examined at all supports while the earthquake loading was applied to the bridge. Existing literature indicates that these rotations provide key information in relation to assessing bridge susceptibility to earthquake damage.

Seismic Design Philosophies and Performance Based Design Criteria

According to building codes, earthquake-resistant structures are intended to withstand the largest earthquake of a certain probability that is likely to occur at their location. This means the loss of life should be minimized by preventing collapse of the buildings for rare earthquakes while the loss of functionality should be limited for more frequent ones.

***Figure:** Snapshot from shake-table video of testing base-isolated (right) and regular (left) building model*

To combat earthquake destruction, the only method available to ancient architects was to build their landmark structures to last, often by making them excessively stiff and strong, like the El Castillo pyramid at Chichen Itza.

Currently, there are several design philosophies in earthquake engineering, making use of experimental results, computer simulations and observations from past earthquakes to offer the required performance for the seismic threat at the site of interest.

These range from appropriately sizing the structure to be strong and ductile enough to survive the shaking with an acceptable damage, to equipping it with base isolation or using structural vibration control technologies to minimize any forces and deformations.

While the former is the method typically applied in most earthquake-resistant structures, important facilities, landmarks and cultural heritage buildings use the more advanced (and expensive) techniques of isolation or control to survive strong shaking with minimal damage. Examples of such applications are the Cathedral of Our Lady of the Angels and the Acropolis Museum.

Trends, Projects

Some of the new state-of-the-art trends and/or projects in the field of earthquake engineering structures are presented below.

Proofed Earthquake Building Material

A German composite construction company developed a proofed earthquake-safe supported core material - based on internal beam/frame constructions. The same principle allows construction of hurricane-safe houses.

Earthquake Shelter

One Japanese construction company has developed a six-foot cubical shelter, presented as an alternative to earthquake-proofing an entire building.

Concurrent Shake-table Testing

Concurrent shake table testing of two or more building models is a vivid, persuasive and effective way to validate earthquake engineering solutions experimentally. Thus, two wooden houses built before adoption of the 1981 Japanese Building Code were moved to E-Defence for testing. The left house was reinforced to enhance its seismic resistance, while the other one was not. These two models were set on E-Defence platform and tested simultaneously.

Combined Vibration Control Solution

Figure: *Close-up of abutment of seismically retrofitted Municipal Services Building in Glendale, CA*

Designed by architect Merrill W. Baird of Glendale, working in collaboration with A. C. Martin Architects of Los Angeles, the Municipal Services Building at 633 East Broadway, Glendale was completed in 1966. Prominently sited at the corner of East Broadway and Glendale Avenue, this civic building serves as a heraldic element of Glendale's civic centre.

In October 2004 Architectural Resources Group (ARG) was contracted by Nabih Youssef & Associates, Structural Engineers, to provide services regarding a historic resource assessment of the building due to a proposed seismic retrofit.

In 2008, the Municipal Services Building of the City of Glendale, California was seismically retrofitted using an innovative combined vibration control solution: the existing elevated building foundation of the building was put on high damping rubber bearings.

Figure: *Seismically retrofitted Municipal Services Building in Glendale, CA*

Steel Plate Shear Walls System

Figure: *Coupled steel plate shear walls, Seattle, Copyright Mehdi Kharrazi*

A steel plate shear wall (SPSW) consists of steel infill plates bounded by a column-beam system. When such infill plates occupy each level within a framed bay of a structure, they constitute a SPSW system. Whereas most earthquake resistant construction methods are adapted from older systems, SPSW was invented entirely to withstand seismic activity.

***Figure:** The Ritz-Carlton/JW Marriott hotel building engaging the advanced steel plate shear walls system, LA*

SPSW behaviour is analogous to a vertical plate girder cantilevered from its base. Similar to plate girders, the SPSW system optimizes component performance by taking advantage of the post-buckling behaviour of the steel infill panels.

The Ritz-Carlton/JW Marriott hotel building, a part of the LA Live development in Los Angeles, California, is the first building in Los Angeles that uses an advanced steel plate shear wall system to resist the lateral loads of strong earthquakes and winds.

Kashiwazaki-Kariwa Nuclear Power Plant is Partially Upgraded

The Kashiwazaki-Kariwa Nuclear Power Plant, the largest nuclear generating station in the world by net electrical power rating, happened to be near the epicenter of the strongest M_w 6.6 July 2007 Chûetsu offshore earthquake. This initiated an extended shutdown for structural inspection which indicated that a greater earthquake-proofing was needed before operation could be resumed.

On May 9, 2009, one unit (Unit 7) was restarted, after the seismic upgrades. The test run had to continue for 50 days. The plant had been completely shut down for almost 22 months following the earthquake.

Seismic Test of Seven-Story Building

Figure: *Seven-story wooden structure prior to shake-table testing, Japan*

A destructive earthquake struck a lone, wooden condominium in Japan. The experiment was webcast live on July 14, 2009 to yield insight on how to make wooden structures stronger and better able to withstand major earthquakes.

The Miki shake at the Hyogo Earthquake Engineering Research Centre is the capstone experiment of the four-year NEESWood project, which receives its primary support from the U.S. National Science Foundation Network for Earthquake Engineering Simulation (NEES) Programme.

"NEESWood aims to develop a new seismic design philosophy that will provide the necessary mechanisms to safely increase the height of wood-frame structures in active seismic zones of the United States, as well as mitigate earthquake damage to low-rise wood-frame structures," said Rosowsky, Department of Civil Engineering at Texas A&M University. This philosophy is based on the application of seismic damping systems for wooden buildings. The systems, which can be installed inside the walls of most wooden buildings, include strong metal frame, bracing and dampers filled with viscous fluid.

Seismic Retrofit

Seismic retrofitting is the modification of existing structures to make them more resistant to seismic activity, ground motion, or soil failure due to earthquakes. With better understanding of seismic demand on structures and with our recent experiences with large earthquakes near urban centres, the need of seismic retrofitting is well acknowledged. Prior to the introduction of modern seismic codes in the late 1960s for developed countries (US, Japan etc.) and late 1970s for many other parts of the world (Turkey, China etc.), many structures were designed without adequate detailing and reinforcement for seismic protection.

In view of the imminent problem, various research work has been carried out. State-of-the-art technical guidelines for seismic assessment, retrofit and rehabilitation have been published around the world - such as the ASCE-SEI 41 and the New Zealand Society for Earthquake Engineering (NZSEE)'s guidelines. These codes must be regularly updated; the 1994 Northridge Earthquake brought to light the brittleness of welded steel frames, for example.

The retrofit techniques outlined here are also applicable for other natural hazards such as tropical cyclones, tornadoes, and severe winds from thunderstorms. Whilst current practice of seismic retrofitting is predominantly concerned with structural improvements to reduce the seismic hazard of using the structures, it is similarly essential to reduce the hazards and losses from non-structural elements. It is also important to keep in mind that there is no such thing as an earthquake-proof structure, although seismic performance can be greatly enhanced through proper initial design or subsequent modifications.

Strategies

Seismic retrofit (or rehabilitation) strategies have been developed in the past few decades following the introduction of new seismic provisions and the availability of advanced materials (e.g. fibre-reinforced polymers (FRP), fibre reinforced concrete and high strength steel). Retrofit strategies are different from retrofit techniques, where the former is the basic approach to achieve an overall retrofit performance objective, such as increasing strength, increasing deformability, reducing deformation demands while the latter is the technical methods to achieve that strategy, for example FRP jacketing.

- Increasing the global capacity (strengthening). This is typically done by the addition of cross braces or new structural walls.
- Reduction of the seismic demand by means of supplementary damping and/or use of base isolation systems.
- Increasing the local capacity of structural elements. This strategy recognises the inherent capacity within the existing structures, and therefore adopt a more cost-effective approach to selectively upgrade local capacity (deformation/ductility, strength or stiffness) of individual structural components.
- Selective weakening retrofit. This is a counter intuitive strategy to change the inelastic mechanism of the structure, whilst recognising the inherent capacity of the structure.
- Allowing sliding connections such as passageway bridges to accommodate additional movement between seismically independent structures.

Performance Objectives

In the past, seismic retrofit was primarily applied to achieve public safety, with engineering solutions limited by economic and political considerations.

However, with the development of Performance based earthquake engineering (PBEE), several levels of performance objectives are gradually recognised:

- Public safety only. The goal is to protect human life, ensuring that the structure will not collapse upon its occupants or passersby, and that the structure can be safely exited. Under severe seismic conditions the structure may be a total economic write-off, requiring tear-down and replacement.
- Structure survivability. The goal is that the structure, while remaining safe for exit, may require extensive repair (but not replacement) before it is generally useful or considered safe for occupation. This is typically the lowest level of retrofit applied to bridges.

- Structure functionality. Primary structure undamaged and the structure is undiminished in utility for its primary application. A high level of retrofit, this ensures that any required repairs are only "cosmetic" - for example, minor cracks in plaster, drywall and stucco. This is the minimum acceptable level of retrofit for hospitals.
- Structure unaffected. This level of retrofit is preferred for historic structures of high cultural significance.

Techniques

Common seismic retrofitting techniques fall into several categories:

Figure: *One of many "earthquake bolts" found throughout period houses in the city of Charleston subsequent to the Charleston earthquake of 1886. They could be tightened and loosened to support the house without having to otherwise demolish the house due to instability. The bolts were directly loosely connected to the supporting frame of the house.*

External Post-tensioning

The use of external post-tensioning for new structural systems have been developed in the past decade. Under the PRESS (Precast Seismic Structural Systems), a large-scale U.S./Japan joint research programme, unbonded post-tensioning high strength steel tendons have been used to achieve a moment-resisting system that has self-centering capacity. An extension of the same idea for seismic retrofitting has been experimentally tested for seismic retrofit of California bridges under a Caltrans research project and for seismic retrofit of non-ductile reinforced concrete frames. Pre-stressing can increase the capacity of structural elements such as beam, column and beam-column joints. It should be noted that external pre-stressing has been used for structural upgrade for gravity/live loading since 1970s.

Base Isolators

Base isolation is a collection of structural elements of a building that should substantially decouple the building's structure from the shaking ground thus protecting the building's integrity and enhancing its seismic performance. This earthquake engineering technology, which is a kind of seismic vibration control, can be applied both to a newly designed building and to seismic upgrading of existing structures. Normally, excavations are made around the building and the building is separated from the foundations. Steel or reinforced concrete beams replace the connections to the foundations, while under these, the isolating pads, or base isolators, replace the material removed. While the base isolation tends to restrict transmission of the ground motion to the building, it also keeps the building positioned properly over the foundation. Careful attention to detail is required where the building interfaces with the ground, especially at entrances, stairways and ramps, to ensure sufficient relative motion of those structural elements.

Supplementary Dampers

Supplementary dampers absorb the energy of motion and convert it to heat, thus "damping" resonant effects in structures that are rigidly attached to the ground. In addition to adding energy dissipation capacity to the structure, supplementary damping can reduce the displacement and acceleration demand within the structures. In some cases, the threat of damage does not come from the initial shock itself, but rather from the periodic resonant motion of the structure that repeated ground motion induces. In the practical sense, supplementary dampers act similarly to Shock absorbers used in automotive suspensions.

Tuned Mass Dampers

Tuned mass dampers (TMD) employ movable weights on some sort of springs. These are typically employed to reduce wind sway in very tall, light buildings. Similar designs may be employed to impart earthquake resistance in eight to ten story buildings that are prone to destructive earthquake induced resonances.

Slosh Tank

A slosh tank is a large tank of fluid placed on an upper floor. During a seismic event, the fluid in this tank will slosh back and forth, but is directed by baffles - partitions that prevent the tank itself becoming resonant; through its mass the water may change or counter the resonant period of the building. Additional kinetic energy can be converted to heat by the baffles and is dissipated through the water - any temperature rise will be insignificant.

Active Control System

Very tall buildings ("skyscrapers"), when built using modern lightweight materials, might sway uncomfortably (but not dangerously) in certain wind conditions. A solution to this problem is to include at some upper story a large mass, constrained, but free to move within a limited range, and moving on some sort of bearing system such as an air cushion or hydraulic film. Hydraulic pistons, powered by electric pumps and accumulators, are actively driven to counter the wind forces and natural resonances. These may also, if properly designed, be effective in controlling excessive motion - with or without applied power - in an earthquake. In general, though, modern steel frame high rise buildings are not as subject to dangerous motion as are medium rise (eight to ten story) buildings, as the resonant period of a tall and massive building is longer than the approximately one second shocks applied by an earthquake.

Adhoc Addition of Structural Support/reinforcement

The most common form of seismic retrofit to lower buildings is adding strength to the existing structure to resist seismic forces. The strengthening may be limited to connections between existing building elements or it may involve adding primary resisting elements such as walls or frames, particularly in the lower stories.

Connections Between Buildings and Their Expansion Additions

Frequently, building additions will not be strongly connected to the existing structure, but simply placed adjacent to it, with only minor continuity in flooring, siding, and roofing. As a result, the addition may have a different resonant period than the original structure, and they may easily detach from one another. The relative motion will then cause the two parts to collide, causing severe structural damage. Seismic modification will either tie the two building components rigidly together so that they behave as a single mass or it will employ dampers to expend the energy from relative motion, with appropriate allowance for this motion, such as increased spacing and sliding bridges between sections.

Exterior Reinforcement of Building

Exterior Concrete Columns

Historic buildings, made of unreinforced masonry, may have culturally important interior detailing or murals that should not be disturbed. In this case, the solution may be to add a number of steel, reinforced concrete, or poststressed concrete columns to the exterior. Careful attention must be paid to the connections with other members such as footings, top plates, and roof trusses.

Infill Shear Trusses

Shown here is an exterior shear reinforcement of a conventional reinforced concrete dormitory building. In this case, there was sufficient vertical strength in the building columns and sufficient shear strength in the lower stories that only limited shear reinforcement was required to make it earthquake resistant for this location near the Hayward fault.

Massive Exterior Structure

In other circumstances, far greater reinforcement is required. In the structure shown at right — a parking garage over shops — the placement, detailing, and painting of the reinforcement becomes itself an architectural embellishment.

Typical Retrofit Scenario & Solution

Soft-story Failure

This collapse mode is known as *soft story collapse.* In many buildings the ground level is designed for different uses than the upper levels. Low rise residential structures may be built over a parking garage which have large doors on one side. Hotels may have a tall ground floors to allow for a grand entrance or ballrooms. Office buildings may have stores in the ground floor which desire continuous windows for display.

Figure: Partial failure due to inadequate shear structure at garage level. Damage in San Francisco due to the Loma Prieta event.

Traditional seismic design assumes that the lower stories of a building are stronger than the upper stories and where this is not the case—if the lower story is less strong than the upper structure—the structure will not respond to earthquakes in the expected fashion. Using modern design methods, it is possible to take a weak story into account. Several failures of this type in one large apartment complex caused most of the fatalities in the 1994 Northridge earthquake.

Typically, where this type of problem is found, the weak story is reinforced to make it stronger than the floors above by adding shear walls or moment frames. Moment frames consisting of inverted U bents are useful in preserving lower story garage access, while a lower cost solution may be to use shear walls or trusses in several locations, which partially reduce the usefulness for automobile parking but still allow the space to be used for other storage.

Beam-column Joint Connections

Beam-column joint connections are a common structural weakness in dealing with seismic retrofitting. Prior to the introduction of modern seismic codes in early 1970s, beam-column joints were typically nonengineered or designed. Laboratory testings have confirmed the seismic vulnerability of these poorly detailed and under-designed connections. Failure of beam-column joint connections can typically lead to catastrophic collapse of a frame-building, as often observed in recent earthquakes.

Figure: *Corner joint steel reinforcement and high tensile strength rods with grouted anti-burst jacket below*

For reinforced concrete beam-column joints - various retrofit solutions have been proposed and tested in the past 20 years. Philosophically, the various seismic retrofit strategies discussed above can be implemented for reinforced concrete joints. Concrete or steel jacketing have been a popular retrofit technique until the advent of composite materials such as Carbon fibre-reinforced polymer (FRP). Composite materials such as carbon FRP and aramic FRP have been extensively tested for use in seismic retrofit with some success. One novel technique includes the use of selective weakening of the beam and added external post-tensioning to the joint in order to achieve flexural hinging in the beam, which is more desirable in terms of seismic design.

Widespread weld failures at beam-column joints of low-to-medium rise steel buildings during the Northridge 1994 earthquake for example, have shown the structural defiencies of these 'modern-designed' post-

1970s welded moment-resisting connections. A subsequent SAC research project has documented, tested and proposed several retrofit solutions for these welded steel moment-resisting connections. Various retrofit solutions have been developed for these welded joints - such as a) weld strengthening and b) addition of steel haunch or 'dog-bone' shape flange.

Following the Northridge earthquake, a number of steel moment -frame buildings were found to have experienced brittle fractures of beam to column connections. Discovery of these unanticipated brittle fractures of framing connections was alarming to engineers and the building industry. Starting in the 1960s, engineers began to regard welded steel moment-frame buildings as being among the most ductile systems contained in the building code. Many engineers believed that steel moment-frame buildings were essentially invulnerable to earthquake induced damage and thought that should damage occur, it would be limited to ductile yielding of members and connections. Observation of damage sustained by buildings in the 1994 Northridge earthquake indicated that contrary to the intended behaviour, in many cases, brittle fractures initiated within the connections at very low levels of plastic demand.

In September, 1994, The SAC joint Venture, AISC, AISI, and NIST jointly convened an international workshop in Los Angeles to coordinate the efforts of various participants and to lay the foundation for systematic investigation and resolution of the problem. In September 1995 the SAC Joint Venture entered into a contractual agreement with FEMA to conduct Phase II of the SAC Steel project. Under Phase II, SAC continued its extensive problem-focused study of the performance of moment resisting steel frames and connections of various configurations, with the ultimate goal of developing seismic design criteria for steel construction. As a result of these studies it is now known that the typical moment resiting connection detail employed in steel moment frame construction prior to the 1994 Northridge earthquake had a number of features that rendered it inherently susceptible to brittle fracture.

Shear Failure Within Floor Diaphragm

Floors in wooden buildings are usually constructed upon relatively deep spans of wood, called joists, covered with a diagonal wood planking or plywood to form a subfloor upon which the finish floor surface is laid. In many structures these are all aligned in the same direction.

To prevent the beams from tipping over onto their side, blocking is used at each end, and for additional stiffness, blocking or diagonal wood or metal bracing may be placed between beams at one or more points in their spans. At the outer edge it is typical to use a single depth of blocking and a perimeter beam overall.

If the blocking or nailing is inadequate, each beam can be laid flat by the shear forces applied to the building. In this position they lack most of their original strength and the structure may further collapse. As part of a retrofit the blocking may be doubled, especially at the outer edges of the building. It may be appropriate to add additional nails between the sill plate of the perimeter wall erected upon the floor diaphragm, although this will require exposing the sill plate by removing interior plaster or exterior siding. As the sill plate may be quite old and dry and substantial nails must be used, it may be necessary to pre-drill a hole for the nail in the old wood to avoid splitting. When the wall is opened for this purpose it may also be appropriate to tie vertical wall elements into the foundation using speciality connectors and bolts glued with epoxy cement into holes drilled in the foundation.

Sliding off Foundation and "Cripple Wall" Failure

***Figure:** House slid off of foundation*

Single or two story wood-frame domestic structures built on a perimeter or slab foundation are relatively safe in an earthquake, but in many structures built before 1950 the sill plate that sits between the concrete foundation and the floor diaphragm (perimeter foundation) or studwall (slab foundation) may not be sufficiently bolted in. Additionally, older attachments (without substantial corrosion-proofing) may have corroded to a point of weakness. A sideways shock can slide the building entirely off of the foundations or slab.

Figure: *Low cripple wall collapse and detachment of structure from concrete stairway*

Often such buildings, especially if constructed on a moderate slope, are erected on a platform connected to a perimeter foundation through low stud-walls called "cripple wall" or *pin-up*. This low wall structure itself may fail in shear or in its connections to itself at the corners, leading to the building moving diagonally and collapsing the low walls.

The likelihood of failure of the pin-up can be reduced by ensuring that the corners are well reinforced in shear and that the shear panels are well connected to each other through the corner posts. This requires structural grade sheet plywood, often treated for rot resistance. This grade of plywood is made without interior unfilled knots and with more, thinner layers than common plywood. New buildings designed to resist earthquakes will typically use OSB (oriented strand board), sometimes with metal joins between panels, and with well attached stucco covering to enhance its performance.

In many modern tract homes, especially those built upon expansive (clay) soil the building is constructed upon a single and relatively thick monolithic slab, kept in one piece by high tensile rods that are stressed after the slab has set. This poststressing places the concrete under compression - a condition under which it is extremely strong in bending and so will not crack under adverse soil conditions.

Multiple Piers in Shallow Pits

Some older low-cost structures are elevated on tapered concrete pylons set into shallow pits, a method frequently used to attach outdoor decks to existing buildings. This is seen in conditions of damp soil, especially in tropical conditions, as it leaves a dry ventilated space under the house, and in far northern conditions of permafrost (frozen mud) as it keeps the building's warmth from destabilizing the ground beneath. During an earthquake, the pylons may tip, spilling the building to the ground. This can be overcome by using deep-bored holes to contain cast-in-place reinforced pylons, which are then secured to the floor panel at the corners of the building. Another technique is to add sufficient diagonal bracing or sections of concrete shear wall between pylons.

Reinforced Concrete Column Burst

Figure: *Jacketed and grouted column on left, unmodified on right*

Reinforced concrete columns typically contain large diameter vertical rebar (reinforcing bars) arranged in a ring, surrounded by lighter-gauge hoops of rebar. Upon analysis of failures due to earthquakes, it has been realized that the weakness was not in the vertical bars, but rather in inadequate strength and quantity of hoops. Once the integrity of the hoops is breached, the vertical rebar can flex outward, stressing the central column of concrete. The concrete then simply crumbles into small pieces, now unconstrained by the surrounding rebar. In new construction a greater amount of hoop-like structures are used.

One simple retrofit is to surround the column with a jacket of steel plates formed and welded into a single cylinder. The space between the jacket and the column is then filled with concrete, a process called grouting. Where soil or structure conditions require such additional modification, additional pilings may be driven near the column base and concrete pads linking the pilings to the pylon are fabricated at or below ground level. In the example shown not all columns needed to be modified to gain sufficient seismic resistance for the conditions expected. (This location is about a mile from the Hayward Fault Zone.)

Reinforced Concrete Wall Burst

Concrete walls are often used at the transition between elevated road fill and overpass structures. The wall is used both to retain the soil and so enable the use of a shorter span and also to transfer the weight of the span directly downward to footings in undisturbed soil. If these walls are inadequate they may crumble under the stress of an earthquake's induced ground motion.

One form of retrofit is to drill numerous holes into the surface of the wall, and secure short L-shaped sections of rebar to the surface of each hole with epoxy adhesive. Additional vertical and horizontal rebar is then secured to the new elements, a form is erected, and an additional layer of concrete is poured. This modification may be combined with additional footings in excavated trenches and additional support ledgers and tie-backs to retain the span on the bounding walls.

Brick Wall Resin and Glass Fibre Reinforcement

Brick building structures have been reinforced with coatings of glass fibre and appropriate resin (epoxy or polyester). In lower floors these may be applied over entire exposed surfaces, while in upper floors this may be confined to narrow areas around window and door openings. This application provides tensile strength that stiffens the wall against bending away from the side with the application. The efficient protection of an entire building requires extensive analysis and engineering to determine the appropriate locations to be treated.

Lift

Where moist or poorly consolidated alluvial soil interfaces in a "beach like" structure against underlying firm material, seismic waves travelling through the alluvium can be amplified, just as are water waves against a sloping beach. In these special conditions, vertical accelerations up to twice the force of gravity have been measured. If a building is not secured to a well-embedded foundation it is possible for the building to be thrust from (or with) its foundations into the air, usually with severe damage upon landing. Even if it is well-founded, higher portions such as upper stories or roof structures or attached structures such as canopies and porches may become detached from the primary structure.

Good practices in modern, earthquake-resistant structures dictate that there be good vertical connections throughout every component of the building, from undisturbed or engineered earth to foundation to sill plate to vertical studs to plate cap through each floor and continuing to the roof structure. Above the foundation and sill plate the connections are typically made using steel strap or sheet stampings, nailed to wood members using special hardened high-shear strength nails, and heavy angle stampings secured with through bolts, using large washers to prevent pull-through. Where inadequate bolts are provided between the sill plates and a foundation in existing construction (or are not trusted due to possible corrosion), special clamp plates may be added, each of which is secured to the foundation using expansion bolts inserted into holes drilled in an exposed face of concrete. Other members must then be secured to the sill plates with additional fittings.

Soil

One of the most difficult retrofits is that required to prevent damage due to soil failure. Soil failure can occur on a slope, a slope failure or landslide, or in a flat area due to liquefaction of water-saturated sand and/or mud. Generally, deep pilings must be driven into stable soil (typically hard mud or sand) or to underlying bedrock or the slope must be stabilized. For buildings built atop previous landslides the practicality of retrofit may be limited by economic factors, as it is not practical to stabilize a large, deep landslide. The likelihood of landslide or soil failure may also depend upon seasonal factors, as the soil may be more stable at the beginning of a wet season than at the beginning of the dry season. Such a "two season" *Mediterranean climate* is seen throughout California.

In some cases, the best that can be done is to reduce the entrance of water runoff from higher, stable elevations by capturing and bypassing through channels or pipes, and to drain water infiltrated directly and from subsurface springs by inserting horizontal perforated tubes. There are numerous locations in California where extensive developments have been built atop archaic landslides, which have not moved in historic times but which (if both water-saturated and shaken by an earthquake) have a high probability of moving *en masse*, carrying entire sections of suburban development to new locations. While the most modern of house structures (well tied to monolithic concrete foundation slabs reinforced with post tensioning cables) may survive such movement largely intact, the building will no longer be in its proper location.

Utility Pipes and Cables: Risks

Natural gas and propane supply pipes to structures often prove especially dangerous during and after earthquakes. Should a building move from its foundation or fall due to cripple wall collapse, the ductile iron pipes transporting the gas within the structure may be broken, typically at the location of threaded joints. The gas may then still be provided to the pressure regulator from higher pressure lines and so continue to flow in substantial quantities; it may then be ignited by a nearby source such as a lit pilot light or arcing electrical connection.

There are two primary methods of automatically restraining the flow of gas after an earthquake, installed on the low pressure side of the regulator, and usually downstream of the gas metre.

- A caged metal ball may be arranged at the edge of an orifice. Upon seismic shock, the ball will roll into the orifice, sealing it to prevent gas flow. The ball may later be reset by the use of an external magnet. This device will respond only to ground motion.
- A flow-sensitive device may be used to close a valve if the flow of gas exceeds a set threshold (very much like an electrical circuit breaker). This device will operate independently of seismic motion, but will not respond to minor leaks which may be caused by an earthquake.

It appears that the most secure configuration would be to use one of each of these devices in series.

Tunnels

Unless the tunnel penetrates a fault likely to slip, the greatest danger to tunnels is a landslide blocking an entrance. Additional protection around the entrance may be applied to divert any falling material (similar as is done to divert snow avalanches) or the slope above the tunnel may be stabilized in some way. Where only small- to medium-sized rocks and boulders are expected to fall, the entire slope may be covered with wire mesh, pinned down to the slope with metal rods. This is also a common modification to highway cuts where appropriate conditions exist.

Underwater Tubes

The safety of underwater tubes is highly dependent upon the soil conditions through which the tunnel was constructed, the materials and reinforcements used, and the maximum predicted earthquake expected, and other factors, some of which may remain unknown under current knowledge.

BART Tube

A tube of particular structural, seismic, economic, and political interest is the BART (Bay Area Rapid Transit) transbay tube. This tube was constructed at the bottom of San Francisco Bay through an innovative process. Rather than pushing a shield through the soft bay mud, the tube was constructed on land in sections. Each section consisted of two inner train tunnels of circular cross section, a central access tunnel of rectangular cross section, and an outer oval shell encompassing the three inner tubes. The intervening space was filled with concrete. At the bottom of the bay a trench was excavated and a flat bed of crushed stone prepared to receive the tube sections. The sections were then floated into place and sunk, then joined with bolted connections to previously-placed sections. An overfill was then placed atop the tube to hold it down. Once completed from San Francisco to Oakland, the tracks and electrical components were installed. The predicted response of the tube during a major earthquake was likened to be as that of a string of (cooked) spaghetti in a bowl of gelatin dessert. To avoid overstressing the tube due to differential movements at each end, a sliding slip joint was included at the San Francisco terminus under the landmark Ferry Building.

The engineers of the construction consortium PBTB (Parsons Brinckerhoff-Tudor-Bechtel) used the best estimates of ground motion available at the time, now known to be insufficient given modern computational analysis methods and geotechnical knowledge. Unexpected settlement of the tube has reduced the amount of slip that can be accommodated without failure.

These factors have resulted in the slip joint being designed too short to ensure survival of the tube under possible (perhaps even likely) large earthquakes in the region. To correct this deficiency the slip joint must be extended to allow for additional movement, a modification expected to be both expensive and technically and logistically difficult. Other retrofits to the BART tube include vibratory consolidation of the tube's overfill to avoid potential liquefying of the overfill, which has now been completed. (Should the overfill fail there is a danger of portions of the tube rising from the bottom, an event which could potentially cause failure of the section connections.)

Bridge Retrofit

Bridges have several failure modes.

Expansion Rockers

Many short bridge spans are statically anchored at one end and attached to rockers at the other. This rocker gives vertical and transverse support while allowing the bridge span to expand and contract with temperature changes. The change in the length of the span is accommodated over a gap in the roadway by comb-like expansion joints. During severe ground motion, the rockers may jump from their tracks or be moved beyond their design limits, causing the bridge to unship from its resting point and then either become misaligned or fail completely. Motion can be constrained by adding ductile or high-strength steel restraints that are friction-clamped to beams and designed to slide under extreme stress while still limiting the motion relative to the anchorage.

Deck Rigidity

Suspension bridges may respond to earthquakes with a side-to-side motion exceeding that which was designed for wind gust response. Such motion can cause fragmentation of the road surface, damage to bearings, and plastic deformation or breakage of components. Devices such as hydraulic dampers or clamped sliding connections and additional diagonal reenforcement may be added.

Figure: *Additional diagonals were inserted under both decks of this bridge*

Lattice Girders, Beams, and ties

Figure: *Obsolete riveted lattice members*

Lattice girders consist of two "I"-beams connected with a criss-cross lattice of flat strap or angle stock. These can be greatly strengthened by replacing the open lattice with plate members. This is usually done in concert with the replacement of hot rivets with bolts.

***Figure:** Bolted plate lattice replacement, forming box members*

Hot Rivets

Many older structures were fabricated by inserting red-hot rivets into pre-drilled holes; the soft rivets are then peened using an air hammer on one side and a bucking bar (an inertial mass) on the head end. As these cool slowly, they are left in an annealed (soft) condition, while the plate, having been hot rolled and quenched during manufacture, remains relatively hard. Under extreme stress the hard plates can shear the soft rivets, resulting in failure of the joint.

The solution is to burn out each rivet with an oxygen torch. The hole is then prepared to a precise diametre with a reamer. A special *locator bolt*, consisting of a head, a shaft matching the reamed hole, and a threaded end is inserted and retained with a nut, then tightened with a wrench. As the bolt has been formed from an appropriate high-strength alloy and has also been heat-treated, it is not subject to either the plastic shear failure typical of hot rivets nor the brittle fracture of ordinary bolts. Any partial failure will be in the plastic flow of the metal secured by the bolt; with proper engineering any such failure should be non-catastrophic.

Fill and Overpass

Elevated roadways are typically built on sections of elevated earth fill connected with bridge-like segments, often supported with vertical columns. If the soil fails where a bridge terminates, the bridge may become disconnected from the rest of the roadway and break away. The retrofit for this is to add additional reinforcement to any supporting wall, or to add deep caissons adjacent to the edge at each end and connect them with a supporting beam under the bridge.

Another failure occurs when the fill at each end moves (through resonant effects) in bulk, in opposite directions. If there is an insufficient founding shelf for the overpass, then it may fall. Additional shelf and ductile stays may be added to attach the overpass to the footings at one or both ends.

The stays, rather than being fixed to the beams, may instead be clamped to them. Under moderate loading, these keep the overpass centred in the gap so that it is less likely to slide off its founding shelf at one end. The ability for the fixed ends to slide, rather than break, will prevent the complete drop of the structure if it should fail to remain on the footings.

Viaducts

Large sections of roadway may consist entirely of viaduct, sections with no connection to the earth other than through vertical columns. When concrete columns are used, the detailing is critical.

Typical failure may be in the toppling of a row of columns due either to soil connection failure or to insufficient cylindrical wrapping with rebar. Both failures were seen in the 1995 Great Hanshin earthquake in Kobe, Japan, where an entire viaduct, centrally supported by a single row of large columns, was laid down to one side. Such columns are reinforced by excavating to the foundation pad, driving additional pilings, and adding a new, larger pad, well connected with rebar alongside or into the column.

A column with insufficient wrapping bar, which is prone to burst and then hinge at the bursting point, may be completely encased in a circular or elliptical jacket of welded steel sheet and grouted as described above.

***Figure:** Cypress Freeway viaduct collapse. Note failure of inadequate anti-burst wrapping and lack of connection between upper and lower vertical elements.*

Sometimes viaducts may fail in the connections between components. This was seen in the failure of the Cypress Freeway in Oakland, California, during the Loma Prieta earthquake. This viaduct was a two-level structure, and the upper portions of the columns were not well connected to the lower portions that supported the lower level; this caused the upper deck to collapse upon the lower deck. Weak connections such as these require additional external jacketing - either through external steel components or by a complete jacket of reinforced concrete, often using stub connections that are glued (using epoxy adhesive) into numerous drilled holes.

These stubs are then connected to additional wrappings, external forms (which may be temporary or permanent) are erected, and additional concrete is poured into the space. Large connected structures similar to the Cypress Viaduct must also be properly analyzed in their entirety using dynamic computer simulations.

Residential Retrofit

Side-to-side forces cause most earthquake damage. Bolting of the mudsill to the foundation and application of plywood to cripple walls are a few basic retrofit techniques which homeowners may apply to wood-framed residential structures to mitigate the effects of seismic activity. The City of San Leandro created guidelines for these procedures, as outlined in the following pamphlet. Public awareness and initiative are critical to the retrofit and preservation of existing building stock, and such efforts as those of the Association of Bay Area Governments are instrumental in providing informational resources to seismically active communities.

Wood Frame Structure

Most houses in North America are wood-framed structures. Wood is one of the best materials for earthquake-resistant construction since it is light weight and more flexible than masonry. It is easy to work with and less expensive than steel, masonry, or concrete. In older homes the most significant weaknesses are the connection from the wood-framed walls to the foundation and the relatively weak "cripple-walls." (Cripple walls are the short wood walls that extend from the top of the foundation to the lowest floor level in houses that have raised floors.)

Adding connections from the base of the wood-framed structure to the foundation is almost always an important part of a seismic retrofit. Bracing the cripple-walls to resist side-to-side forces is essential in houses with cripple walls; bracing is usually done with plywood. Oriented strand board (OSB) does not perform as consistently as plywood, and is not the favoured choice of retrofit designers or installers.

Retrofit methods in older woodframe structures may consist of the following, and other methods not described here.

- The lowest plate rails of walls (usually called "mudsills" or "foundation sills" in North America) are bolted to a continuous foundation, or secured with rigid metal connectors bolted to the foundation so as to resist side-to-side forces.
- *Cripple walls* are braced with plywood.
- Selected vertical elements (typically the posts at the ends of plywood wall bracing panels) are connected to the foundation. These connections are intended to prevent the braced walls from rocking up and down when subjected to back-and-forth forces at the top of the braced walls, not to resist the wall or house "jumping" off the foundation (which almost never occurs).
- In two story buildings using "platform framing" (sometimes called "western" style construction, where walls are progressively erected upon the lower story's upper diaphragm,

unlike "eastern" or *balloon framing*), the upper walls are connected to the lower walls with tension elements. In some cases, connections may be extended vertically to include retention of certain roof elements. This sort of strengthening is usually very costly with respect to the strength gained.

- Vertical posts are secured to the beams or other members they support. This is particularly important where loss of support would lead to collapse of a segment of a building. Connections from posts to beams cannot resist appreciable side-to-side forces; it is much more important to strengthen around the perimeter of a building (bracing the cripple-walls and supplementing foundation-to-wood-framing connections) than it is to reinforce post-to-beam connections.

Wooden framing is efficient when combined with masonry, if the structure is properly designed. In Turkey, the traditional houses (bagdadi) are made with this technology. In El Salvador, wood and bamboo are used for residential construction.

Reinforced and Unreinforced Masonry

In many parts of developing countries such as Pakistan, Iran and China, unreinforced or in some cases reinforced masonry is the predominantly form of structures for rural residential and dwelling. Masonry was also a common construction form in the early part of the 20th century, which implies that a substantial number of these at-risk masonry structures would have significant heritage value.

Masonry walls that are not reinforced are especially hazardous. Such structures may be more appropriate for replacement than retrofit, but if the walls are the principal load bearing elements in structures of modest size they may be appropriately reinforced. It is especially important that floor and ceiling beams be securely attached to the walls. Additional vertical supports in the form of steel or reinforced concrete may be added.

In the western United States, much of what is seen as masonry is actually brick or stone veneer. Current construction rules dictate the amount of *tie back* required, which consist of metal straps secured to vertical structural elements. These straps extend into mortar courses, securing the veneer to the primary structure. Older structures may not secure this sufficiently for seismic safety. A weakly secured veneer in a house interior (sometimes used to face a fireplace from floor to ceiling) can be especially dangerous to occupants. Older masonry chimneys are also dangerous if they have substantial vertical extension above the roof. These are prone to breakage at the roofline and may fall into the house in a single large piece.

For retrofit, additional supports may be added; however, it is extremely expensive to strengthen an existing masonry chimney to conform with contemporary design standards. It is best to simply remove the extension and replace it with lighter materials, with special metal flue replacing the flue tile and a wood structure replacing the masonry.

This may be matched against existing brickwork by using very thin veneer (similar to a tile, but with the appearance of a brick).

3

Seismic Design of Reinforced Concrete Bridges

Concrete is a composite material composed of coarse granular material (the aggregate or filler) embedded in a hard matrix of material (the cement or binder) that fills the space among the aggregate particles and glues them together.

Concrete is widely used for making architectural structures, foundations, brick/block walls, pavements, bridges/overpasses, highways, runways, parking structures, dams, pools/reservoirs, pipes, footings for gates, fences and poles and even boats.

Famous concrete structures include the Hoover Dam, the Panama Canal and the Roman Pantheon.

Concrete technology was known by the Ancient Romans and was widely used within the Roman Empire—the Colosseum is largely built of concrete and the concrete dome of the Pantheon is the world's largest. After the Empire passed, use of concrete became scarce until the technology was re-pioneered in the mid-18th century.

History

The word concrete comes from the Latin word "*concretus*" (meaning compact or condensed), the perfect passive participle of "*concrescere*", from "*con* " (together) and "*crescere*" (to grow).

Perhaps the earliest known occurrence of cement was twelve million years ago, when a natural deposit formed after an occurrence of oil shale naturally combusted while adjacent to a bed of limestone. These ancient deposits were investigated in the 1960s and 1970s.

On a human time-scale, lime mortars were used in Greece, Crete, and Cyprus in 800 BC. The Assyrian Jerwan Aqueduct (688 BC) made use of fully waterproof concrete. German archaeologist Heinrich Schliemann found concrete floors, which were made of lime and pebbles, in the royal palace of Tiryns, Greece, which dates roughly to 1400-1200 BC. Concrete was used for construction in many ancient structures.

The Romans used concrete extensively from 300 BC to 476 AD, a span of more than seven hundred years. During the Roman Empire, Roman concrete (or *opus caementicium*) was made from quicklime, pozzolana and an aggregate of pumice. Its widespread use in many Roman structures, a key event in the history of architecture termed the Roman Architectural Revolution, freed Roman construction from the restrictions of stone and brick material and allowed for revolutionary new designs in terms of both structural complexity and dimension.

Concrete, as the Romans knew it, was a new and revolutionary material. Laid in the shape of arches, vaults and domes, it quickly hardened into a rigid mass, free from many of the internal thrusts and strains that troubled the builders of similar structures in stone or brick.

Modern tests show that *opus caementicium* had as much compressive strength as modern Portland-cement concrete (ca. 200 kg/cm^2). However, due to the absence of reinforcement, its tensile strength was far lower than modern reinforced concrete, and its mode of application was also different:

Modern structural concrete differs from Roman concrete in two important details. First, its mix consistency is fluid and homogeneous, allowing it to be poured into forms rather than requiring hand-layering together with the placement of aggregate, which, in Roman practice, often consisted of rubble. Second, integral reinforcing steel gives modern concrete assemblies great strength in tension, whereas Roman concrete could depend only upon the strength of the concrete bonding to resist tension.

The widespread use of concrete in many Roman structures has ensured that many survive to the present day. The Baths of Caracalla in Rome are just one example. Many Roman aqueducts and bridges have masonry cladding on a concrete core, as does the dome of the Pantheon.

After the Roman Empire, the use of burning lime and pozzolana was greatly reduced until the technique was all but forgotten between 500 AD and the 1300s. Between the 1300s until the mid-1700s, the use of cement gradually returned. The *Canal du Midi* was built using concrete in 1670, and there are concrete structures in Finland that date from the 16th century.

Perhaps the greatest driver behind the modern usage of concrete was the third Eddystone Lighthouse in Devon, England. To create this structure, between 1756 and 1793, British engineer John Smeaton pioneered the use of hydraulic lime in concrete, using pebbles and powdered brick as aggregate.

A method for producing Portland cement was patented by Joseph Aspdin on 1824.

Reinforced concrete was invented in 1849 by Joseph Monier. In 1889 the first concrete reinforced bridge was built, and the first large concrete dams were built in 1936, Hoover Dam and Grand Coulee Dam.

Ancient Additives

Concrete additives have been used since Roman and Egyptian times, when it was discovered that adding volcanic ash to the mix allowed it to set underwater. Similarly, the Romans knew that adding horse hair made concrete less liable to crack while it hardened, and adding blood made it more frost-resistant.

Modern Additives

In modern times, researchers have experimented with the addition of other materials to create concrete with improved properties, such as higher strength or electrical conductivity.

Impact of Modern Concrete Use

The usage of concrete worldwide is twice as much as steel, wood, plastics, and aluminium combined. Concrete's use in the modern world is only exceeded by the usage of naturally occurring water.

Concrete is also the basis of a large commercial industry. In the United States alone, concrete production is a $30 billion per year industry, considering only the value of the ready-mixed concrete sold each year. Given the size of the concrete industry, and the fundamental way concrete is used to shape the infrastructure of the modern world, it is difficult to overstate the role this material plays today.

Composition of Concrete

There are many types of concrete available, created by varying the proportions of the main ingredients below. In this way or by substitution for the cementitious and aggregate phases, the finished product can be tailored to its application with varying strength, density, or chemical and thermal resistance properties.

"Aggregate" consists of large chunks of material in a concrete mix, generally a coarse gravel or crushed rocks such as limestone, or granite, along with finer materials such as sand.

"Cement", commonly Portland cement, and other cementitious materials such as fly ash and slag cement, serve as a binder or a Ligare for the aggregate.

Water is then mixed with this dry composite, which produces a semi-liquid that workers can shape (typically by pouring it into a form). The concrete solidifies and hardens to rock-hard strength through a chemical process called hydration. The water reacts with the cement, which bonds the other components together, creating a robust stone-like material.

"Chemical admixtures" are added to achieve varied properties. These ingredients may speed or slow down the rate at which the concrete hardens, and impart many other useful properties.

"Reinforcements" are often added to concrete. Concrete can be formulated with high compressive strength, but always has lower tensile strength. For this reason it is usually reinforced with materials that are strong in tension (often steel).

"Mineral admixtures" are becoming more popular in recent decades. The use of recycled materials as concrete ingredients has been gaining popularity because of increasingly stringent environmental legislation, and the discovery that such materials often have complimentary and valuable properties.

The most conspicuous of these are fly ash, a by-product of coal-fired power plants, and silica fume, a byproduct of industrial electric arc furnaces.

The use of these materials in concrete reduces the amount of resources required as the ash and fume acts as a cement replacement. This displaces some cement production, an energetically expensive and environmentally problematic process, while reducing the amount of industrial waste that must be disposed of.

The *mix design* depends on the type of structure being built, how the concrete is mixed and delivered, and how it is placed to form the structure.

Cement

Figure: *Blocks of concrete in Belo Horizonte, Brazil.*

Portland cement is the most common type of cement in general usage. It is a basic ingredient of concrete, mortar and plaster. English masonry worker Joseph Aspdin patented Portland cement in 1824. It was named because of the similarity of its colour to Portland limestone, quarried from the English Isle of Portland and used extensively in London architecture.

It consists of a mixture of oxides of calcium, silicon and aluminium. Portland cement and similar materials are made by heating limestone (a source of calcium) with clay and grinding this product (called *clinker*) with a source of sulphate (most commonly gypsum).

In modern cement kilns many advanced features are used to lower the fuel consumption per ton of clinker produced. Cement kilns are extremely large, complex, and inherently dirty industrial installations, and have many undesirable emissions. Of the various ingredients used in concrete the cement is the most energetically expensive. Even complex and efficient kilns require 3.3 to 3.6 gigajoules of energy to produce a ton of clinker and then grind it into cement. Many kilns can be fuelled with difficult to dispose of wastes, the most common being used tires. The extremely high temperatures and long periods of time at those temperatures allows cement kilns to efficiently and completely burn even difficult to use fuels.

In recent years, alternatives have been developed to help replace cement. Products such as PLC (Portland Limestone Cement), which incorporate limestone into the mix, are being tested. This is due to cement production being one of the largest producers of global greenhouse gas emissions (about 5 to 10%).

Water

Combining water with a cementitious material forms a cement paste by the process of hydration. The cement paste glues the aggregate together, fills voids within it, and makes it flow more freely.

A lower water to cement ratio yields a stronger, more durable concrete, while more water gives a freer-flowing concrete with a higher slump. Impure water used to make concrete can cause problems when setting or in causing premature failure of the structure.

Hydration involves many different reactions, often occurring at the same time. As the reactions proceed, the products of the cement hydration process gradually bond together the individual sand and gravel particles and other components of the concrete, to form a solid mass.

Reaction:

Cement chemist notation: $C_3S + H \rightarrow C\text{-}S\text{-}H + CH$

Standard notation: $Ca_3SiO_5 + H_2O \rightarrow (CaO)\cdot(SiO_2)\cdot(H_2O)(gel) + Ca(OH)_2$

Balanced: $2Ca_3SiO_5 + 7H_2O \rightarrow 3(CaO)\cdot 2(SiO_2)\cdot 4(H_2O)(gel) + 3Ca(OH)_2$

Aggregates

Fine and coarse aggregates make up the bulk of a concrete mixture. Sand, natural gravel and crushed stone are used mainly for this purpose. Recycled aggregates (from construction, demolition and excavation waste) are increasingly used as partial replacements of natural aggregates, while a number of manufactured aggregates, including air-cooled blast furnace slag and bottom ash are also permitted.

The presence of aggregate greatly increases the durability of concrete above that of cement, which is a brittle material in its pure state. Thus concrete is a true composite material. Redistribution of aggregates after compaction often creates inhomogeneity due to the influence of vibration. This can lead to strength gradients.

Decorative stones such as quartzite, small river stones or crushed glass are sometimes added to the surface of concrete for a decorative "exposed aggregate" finish, popular among landscape designers.

In addition to being decorative, exposed aggregate adds robustness to a concrete driveway.

Reinforcement

Concrete is strong in compression, as the aggregate efficiently carries the compression load. However, it is weak in tension as the cement holding the aggregate in place can crack, allowing the structure to fail. Reinforced concrete adds either steel reinforcing bars, steel fibres, glass fibre, or plastic fibre to carry tensile loads.

Chemical Admixtures

Chemical admixtures are materials in the form of powder or fluids that are added to the concrete to give it certain characteristics not obtainable with plain concrete mixes. In normal use, admixture dosages are less than 5% by mass of cement and are added to the concrete at the time of batching/mixing. The common types of admixtures are as follows.

- Accelerators speed up the hydration (hardening) of the concrete. Typical materials used are $CaCl_2$, $Ca(NO_3)_2$ and $NaNO_3$. However, use of chlorides may cause corrosion in steel reinforcing and is prohibited in some countries, so that nitrates may be favoured.
- Retarders slow the hydration of concrete and are used in large or difficult pours where partial setting before the pour is complete is undesirable. Typical polyol retarders are sugar, sucrose, sodium gluconate, glucose, citric acid, and tartaric acid.

- Air entrainments add and entrain tiny air bubbles in the concrete, which reduces damage during freeze-thaw cycles, increasing durability. However, entrained air entails a trade off with strength, as each 1% of air may decrease compressive strength 5%.
- Plasticizers increase the workability of plastic or "fresh" concrete, allowing it be placed more easily, with less consolidating effort. A typical plasticizer is lignosulfonate. Plasticizers can be used to reduce the water content of a concrete while maintaining workability and are sometimes called *water-reducers* due to this use. Such treatment improves its strength and durability characteristics. Superplasticizers (also called *high-range water-reducers*) are a class of plasticizers that have fewer deleterious effects and can be used to increase workability more than is practical with traditional plasticizers. Compounds used as superplasticizers include sulphonated naphthalene formaldehyde condensate, sulphonated melamine formaldehyde condensate, acetone formaldehyde condensate and polycarboxylate ethers.
- Pigments can be used to change the colour of concrete, for aesthetics.
- Corrosion inhibitors are used to minimize the corrosion of steel and steel bars in concrete.
- Bonding agents are used to create a bond between old and new concrete (typically a type of polymer) .
- Pumping aids improve pumpability, thicken the paste and reduce separation and bleeding.

Mineral Admixtures and Blended Cements

There are inorganic materials that also have pozzolanic or latent hydraulic properties. These very fine-grained materials are added to the concrete mix to improve the properties of concrete (mineral admixtures), or as a replacement for Portland cement (blended cements).

- Fly ash: A by-product of coal-fired electric generating plants, it is used to partially replace Portland cement (by up to 60% by mass). The properties of fly ash depend on the type of coal burnt. In general, siliceous fly ash is pozzolanic, while calcareous fly ash has latent hydraulic properties.

- Ground granulated blast furnace slag (GGBFS or GGBS): A by-product of steel production is used to partially replace Portland cement (by up to 80% by mass). It has latent hydraulic properties.
- Silica fume: A by-product of the production of silicon and ferrosilicon alloys. Silica fume is similar to fly ash, but has a particle size 100 times smaller. This results in a higher surface to volume ratio and a much faster pozzolanic reaction. Silica fume is used to increase strength and durability of concrete, but generally requires the use of superplasticizers for workability.
- High reactivity Metakaolin (HRM): Metakaolin produces concrete with strength and durability similar to concrete made with silica fume. While silica fume is usually dark gray or black in colour, high-reactivity metakaolin is usually bright white in colour, making it the preferred choice for architectural concrete where appearance is important.

Concrete Production

Figure: *Concrete plant facility (background) with concrete delivery trucks.*

Concrete production is the process of mixing together the various ingredients—water, aggregate, cement, and any additives—to produce concrete. Concrete production is time-sensitive. Once the ingredients are mixed, workers must put the concrete in place before it hardens.

There is a wide variety of equipment for processing concrete—from hand tools to heavy industrial machinery. Whichever equipment builders use however, the objective is to produce the desired building material—ingredients must be properly mixed, placed, shaped, and retained within time constraints. Once the mix is where it should be, the curing process must be controlled to ensure the concrete attains desired attributes. During concrete preparation, various technical details may affect the quality and nature of the product.

When initially mixed, Portland cement and water rapidly form a gel of tangled chains of interlocking crystals, and components of the gel continue to react over time. Initially the gel is fluid, which improves workability and aids in placement of the material, but as the concrete sets, the chains of crystals join into a rigid structure, counteracting the fluidity of the gel and fixing the particles of aggregate in place. During curing, the cement continues to react with the residual water in a process of hydration. In properly formulated concrete, once this curing process has terminated the product has the desired physical and chemical properties. Among the qualities typically desired, are mechanical strength, low moisture permeability, and chemical and volumetric stability.

Mixing Concrete

Thorough mixing is essential for the production of uniform, high quality concrete. For this reason equipment and methods should be capable of effectively mixing concrete materials containing the largest specified aggregate to produce *uniform mixtures* of the lowest slump practical for the work.

Separate paste mixing has shown that the mixing of cement and water into a paste before combining these materials with aggregates can increase the compressive strength of the resulting concrete. The paste is generally mixed in a *high-speed*, shear-type mixer at a w/cm (water to cement ratio) of 0.30 to 0.45 by mass. The cement paste premix may include admixtures such as accelerators or retarders, superplasticizers, pigments, or silica fume. The premixed paste is then blended with aggregates and any remaining batch water and final mixing is completed in conventional concrete mixing equipment.

High-energy mixed (HEM) concrete is produced by means of high-speed mixing of cement, water and sand with net specific energy consumption of at least 5 kilojoules per kilogram of the mix. A plasticizer or a superplasticizer is then added to the activated mixture, which can later be mixed with aggregates in a conventional concrete mixer. In this process, sand provides dissipation of energy and creates high-shear conditions on the surface of cement particles.

This results in the full volume of water interacting with cement. The liquid activated mixture can be used by itself or foamed (expanded) for lightweight concrete. HEM concrete hardens in low and subzero temperature conditions and possesses an increased volume of gel, which drastically reduces capillarity in solid and porous materials.

Workability

Figure: *Pouring and smoothing out concrete at Palisades Park in Washington DC.*

Workability is the ability of a fresh (plastic) concrete mix to fill the form/mold properly with the desired work (vibration) and without reducing the concrete's quality. Workability depends on water content, aggregate (shape and size distribution), cementitious content and age (level of hydration) and can be modified by adding chemical admixtures, like superplasticizer. Raising the water content or adding chemical admixtures increases concrete workability. Excessive water leads to increased bleeding (surface water) and/or segregation of aggregates (when the cement and aggregates start to separate), with the resulting concrete having reduced quality. The use of an aggregate with an undesirable gradation can result in a very harsh mix design with a very low slump, which cannot be readily made more workable by addition of reasonable amounts of water.

Workability can be measured by the concrete slump test, a simplistic measure of the plasticity of a fresh batch of concrete following the ASTM C 143 or EN 12350-2 test standards. Slump is normally measured by filling an "Abrams cone" with a sample from a fresh batch of concrete. The cone is placed with the wide end down onto a level, non-absorptive surface. It is then filled in three layers of equal volume, with each layer being tamped with a steel rod to consolidate the layer. When the cone is carefully lifted off, the enclosed material slumps a certain amount due to gravity. A relatively dry sample slumps very little, having a slump value of one or two inches (25 or 50 mm) out of one foot (305 mm). A relatively wet concrete sample may slump as much as eight inches. Workability can also be measured by using the flow table test.

Slump can be increased by addition of chemical admixtures such as plasticizer or superplasticizer without changing the water-cement ratio. Some other admixtures, especially air-entraining admixture, can increase the slump of a mix.

High-flow concrete, like self-consolidating concrete, is tested by other flow-measuring methods. One of these methods includes placing the cone on the narrow end and observing how the mix flows through the cone while it is gradually lifted.

After mixing, concrete is a fluid and can be pumped to the location where needed.

Curing

In all but the least critical applications, care must be taken to properly *cure* concrete, to achieve best strength and hardness. This happens after the concrete has been placed. Cement requires a moist, controlled environment to gain strength and harden fully. The cement paste hardens over time, initially setting and becoming rigid though very weak and gaining in strength in the weeks following. In around 4 weeks, typically over 90% of the final strength is reached, though strengthening may continue for decades. The conversion of calcium hydroxide in the concrete into calcium carbonate from absorption of

CO_2 over several decades further strengthen the concrete and make it more resilient to damage. However, this reaction, called carbonation, lowers the pH of the cement pore solution and can cause the reinforcement bars to corrode.

***Figure:** A concrete slab ponded while curing.*

Hydration and hardening of concrete during the first three days is critical. Abnormally fast drying and shrinkage due to factors such as evaporation from wind during placement may lead to increased tensile stresses at a time when it has not yet gained sufficient strength, resulting in greater shrinkage cracking. The early strength of the concrete can be increased if it is kept damp during the curing process. Minimizing stress prior to curing minimizes cracking. High-early-strength concrete is designed to hydrate faster, often by increased use of cement that increases shrinkage and cracking. The strength of concrete changes (increases) for up to three years. It depends on cross-section dimension of elements and conditions of structure exploitation.

During this period concrete must be kept under controlled temperature and humid atmosphere. In practice, this is achieved by spraying or ponding the concrete surface with water, thereby protecting the concrete mass from ill effects of ambient conditions. The pictures to the right show two of many ways to achieve this, ponding – submerging setting concrete in water and wrapping in plastic to contain the water in the mix. Additional common curing methods include wet burlap and/or plastic sheeting covering the fresh concrete, or by spraying on a water-impermeable temporary curing membrane.

Properly curing concrete leads to increased strength and lower permeability and avoids cracking where the surface dries out prematurely. Care must also be taken to avoid freezing or overheating due to the exothermic setting of cement. Improper curing can cause scaling, reduced strength, poor abrasion resistance and cracking.

Properties

Concrete has relatively high compressive strength, but much lower tensile strength. For this reason it is usually reinforced with materials that are strong in tension (often steel). The elasticity of concrete is relatively constant at low stress levels but starts decreasing at higher stress levels as matrix cracking develops. Concrete has a very low coefficient of thermal expansion and shrinks as it matures. All concrete structures crack to some extent, due to shrinkage and tension. Concrete that is subjected to long-duration forces is prone to creep.

Tests can be made to ensure the properties of concrete correspond to specifications for the application.

Different mixes of concrete ingredients produce different strengths, which are measured in psi or MPa.

Different strengths of concrete are used for different purposes. Very low-strength (2000 psi or less) concrete may be used when the concrete must be lightweight. Lightweight concrete is often achieved by adding air, foams, or lightweight aggregates, with the side effect that the strength is reduced. For most routine uses, 3000 psi to 4000 psi concrete is often used. 5000 psi concrete is readily commercially available as a more durable, although more expensive, option. 5000 psi concrete is often used for larger civil projects.

Strengths above 5000 psi are often used for specific building elements. For example, the lower floor columns of high-rise concrete buildings may use concrete of 12,000 psi or more, to keep the size of the columns small. Bridges may use long beams of 10,000 psi concrete to lower the number of spans required. Occasionally, other structural needs may require high strength concrete. If a structure must be very rigid, concrete of very high strength may be specified, even much stronger than is required to bear the service loads. Strengths as high as 19,000 psi have been used commercially for these reasons.

Figure: *Compression testing of a concrete cylinder*

Imperial Strength	***Metric Equivalent***
2000 psi	14 MPa
2500 psi	18 MPa
3000 psi	20 MPa
3500 psi	25 MPa
4000 psi	30 MPa
5000 psi	35 MPa
6000 psi	40 MPa
7000 psi	50 MPa
8000 psi	55 MPa
10000 psi	70 MPa

Concrete Degradation

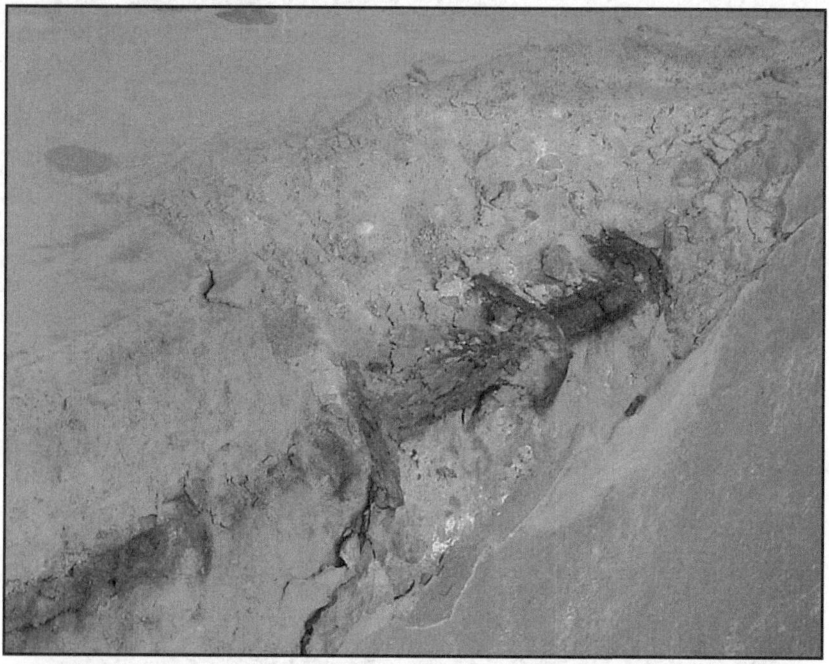

Figure: Concrete spalling caused by the corrosion of rebar

Concrete can be damaged by many processes, such as the expansion of corrosion products of the steel reinforcement bars, freezing of trapped water, fire or radiant heat, aggregate expansion, sea water effects, bacterial corrosion, leaching, erosion by fast-flowing water, physical damage and chemical damage (from carbonatation, chlorides, sulphates and distillate water).

Environmental and Health

The manufacture and use of concrete produce a wide range of environmental and social consequences. Some are harmful, some welcome, and some both, depending on circumstances. A major component of concrete is cement, which similarly exerts environmental and social effects.

The cement industry is one of the three primary producers of carbon dioxide, a major greenhouse gas. The other two are the energy production and transportation industries.

Concrete is used to create hard surfaces that contribute to surface runoff, which can cause heavy soil erosion, water pollution, and flooding, but conversely can be used to divert, dam, and control flooding.

Concrete is a primary contributor to the urban heat island effect, though less so than asphalt.

Workers who cut, grind or polish concrete are at risk of inhaling airborne silica, which can lead to silicosis. Concrete dust released by building demolition and natural disasters can be a major source of dangerous air pollution.

The presence of some substances in concrete, including useful and unwanted additives, can cause health concerns due to toxicity and radioactivity. Wet concrete is highly alkaline and must be handled with proper protective equipment.

Figure: *Recycled crushed concrete, to be reused as granular fill, is loaded into a semi-dump truck.*

Concrete Recycling

Concrete recycling is an increasingly common method of disposing of concrete structures. Concrete debris was once routinely shipped to landfills for disposal, but recycling is increasing due to improved environmental awareness, governmental laws and economic benefits.

Concrete, which must be free of trash, wood, paper and other such materials, is collected from demolition sites and put through a crushing machine, often along with asphalt, bricks and rocks. Reinforced concrete contains rebar and other metallic reinforcements, which are removed with magnets and recycled elsewhere. The remaining aggregate chunks are sorted by size. Larger chunks may go through the crusher again. Smaller pieces of concrete are used as gravel for new construction projects. Aggregate base gravel is laid down as the lowest layer in a road, with fresh concrete or asphalt placed over it. Crushed recycled concrete can sometimes be used as the dry aggregate for brand new concrete if it is free of contaminants, though the use of recycled concrete limits strength and is not allowed in many jurisdictions. On 3 March 1983, a government funded research team (the VIRL research.codep) approximated that almost 17% of worldwide landfill was by-products of concrete based waste.

Use of Concrete in Infrastructure

Figure: *Aerial photo of reconstruction at Taum Sauk (Missouri) pumped storage facility in late November, 2009. After the original reservoir failed, the new reservoir is made of roller compacted concrete.*

Mass Concrete Structures

Large concrete structures such as dams, navigation locks, large mat foundations, and large breakwaters generate excessive heat during cement hydration and associated expansion. To mitigate these effects post-cooling is commonly applied during construction. An early example at Hoover Dam, installed a network of pipes between vertical concrete placements to circulate cooling water during the curing process to avoid damaging overheating. Similar systems are still used; depending on volume of the pour, the concrete mix used, and ambient air temperature, the cooling process may last for many months after the concrete is placed. Various methods also are used to pre-cool the concrete mix in mass concrete structures.

Another approach to mass concrete structures that is becoming more widespread is the use of roller-compacted concrete, which uses much lower levels of cement and water than traditional concrete mixtures, and is generally not poured into place. Instead it is placed in thick layers as a semi-dry material and compacted into a dense, strong mass with rolling compactors. Because it uses less cementitious material, Roller Compacted Concrete has a much lower cooling requirement than traditional concrete.

Prestressed Concrete Structures

Figure: *40-foot cacti decorate a sound/retaining wall in Scottsdale, Arizona*

Prestressed concrete is a form of reinforced concrete that builds in compressive stresses during construction to oppose those found when in use. This can greatly reduce the weight of beams or slabs, by better distributing the stresses in the structure to make optimal use of the reinforcement. For example a horizontal beam tends to sag. Prestressed reinforcement along the bottom of the beam counteracts this. In pre-tensioned concrete, the prestressing is achieved by using steel or polymer tendons or bars that are subjected to a tensile force prior to casting, or for post-tensioned concrete, after casting.

Concrete Textures

When one thinks of concrete, the image of a dull, gray concrete wall often comes to mind. With the use of form liner, concrete can be cast and molded into different textures and used for decorative concrete applications. Sound/retaining walls, bridges, office buildings and more serve as the optimal canvases for concrete art. For example, the Pima Freeway/Loop 101 retaining and sound walls in Scottsdale, Arizona, feature desert flora and fauna, a 67-foot (20 m) lizard and 40-foot (12 m) cacti along the 8-mile (13 km) stretch. The project, titled "The Path Most Travelled," is one example of how concrete can be shaped using elastomeric form liner.

Building with Concrete

Concrete is one of the most durable building materials. It provides superior fire resistance, compared with wooden construction and can gain strength over time. Structures made of concrete can have a long service life. Concrete is the most widely used construction material in the world with annual consumption estimated at between 21 and 31 billion tonnes.

Figure: *The Buffalo City Court Building in Buffalo, NY.*

Concrete is used more than any other man-made material in the world. As of 2006, about 7.5 billion cubic metres of concrete are made each year—more than one cubic metre for every person on Earth.

Concrete powers a US$35 billion industry, employing more than two million workers in the United States alone. More than 55,000 miles (89,000 km) of highways in the United States are paved with this material. Reinforced concrete, prestressed concrete and precast concrete are the most widely used types of concrete functional extensions in modern days.

Energy Efficiency

Energy requirements for transportation of concrete are low because it is produced locally from local resources, typically manufactured within 100 kilometers of the job site. Similarly, relatively little energy is used in producing and combining the raw materials (although large amounts of CO_2 are produced by the chemical reactions in cement manufacture). The overall embodied energy of concrete is therefore lower than for most structural materials other than wood.

Once in place, concrete offers significant energy efficiency over the lifetime of a building. Concrete walls leak air far less than those made of wood-frames. Air leakage accounts for a large percentage of energy loss from a home.

The thermal mass properties of concrete increase the efficiency of both residential and commercial buildings. By storing and releasing the energy needed for heating or cooling, concrete's thermal mass delivers year-round benefits by reducing temperature swings inside and minimizing heating and cooling costs. While insulation reduces energy loss through the building envelope, thermal mass uses walls to store and release energy. Modern concrete wall systems use both external insulation and thermal mass to create an energy-efficient building. Insulating Concrete Forms (ICFs) are hollow blocks or panels made of either insulating foam or rastra that are stacked to form the shape of the walls of a building and then filled with reinforced concrete to create the structure.

Pervious Concrete

Pervious concrete is a mix of specially graded coarse aggregate, cement, water and little-to-no fine aggregates. This concrete is also known as "no-fines" or porous concrete. Mixing the ingredients in a carefully controlled process creates a paste that coats and bonds the aggregate particles. The hardened concrete contains interconnected air voids totalling approximately 15 to 25 percent. Water runs through the voids in the pavement to the soil underneath. Air entrainment admixtures are often used in freeze-thaw climates to minimize the possibility of frost damage.

Fire Safety

Figure: *A modern building: Boston City Hall (completed 1968) is constructed largely of concrete, both precast and poured in place. Of Brutalist architecture, it was voted "The World's Ugliest Building" in 2008.*

Concrete buildings are more resistant to fire than those constructed using steel frames, since concrete has lower heat conductivity than steel and therefore could last longer under the same fire conditions. Concrete is sometimes used as a fire protection for steel frames, for the same effect as above. Concrete as a fire shield, for example Fondu fyre, can also be used in extreme environments like a missile launch pad.

Options for non-combustible construction include floors, ceilings and roofs made of cast-in-place and hollow-core precast concrete. For walls, concrete masonry technology and Insulating Concrete Forms (ICFs) are additional options. ICFs are hollow blocks or panels made of fire-proof insulating foam that are stacked to form the shape of the walls of a building and then filled with reinforced concrete to create the structure.

Concrete also provides a good resistance against external forces applied such as high winds, hurricanes, tornadoes due to its lateral stiffness that results in minimal horizontal movement. However this stiffness can work against certain types of structure being constructed in concrete, particularly where a relatively higher flexing structure is require to resist more extreme forces.

Earthquake Safety

As discussed above, concrete is very strong in compression, but weak in tension. Larger earthquakes can generate very large shear loads on structures. These shear loads subject the structure to both tensile and compressional loads. Concrete structures without reinforcing, like other unreinforced masonry structures, can fail during severe earthquake shaking. Unreinforced masonry structures constitute one of the largest earthquake risks globally. These risks can be reduced through seismic retrofitting of at-risk buildings, (e.g. School buildings in Istanbul, Turkey).

Useful Life

Figure: *The Paulins Kill Viaduct was completed in 1910.*

Concrete can be viewed as a form of artificial sedimentary rock. As a type of mineral, the compounds of which it is composed are extremely stable. Many concrete structures are built with an expected lifetime of approximately 100 years, but researchers have suggested that adding silica fume could extend the useful life of bridges and other concrete uses to as long as 16,000 years. Coatings are also available to protect concrete from damage, and extend the useful life. Epoxy coatings may only be applied to interior surfaces, though, as they would otherwise trap moisture in the concrete.

Self healing concrete has been developed, which can also last longer than traditional concrete.

Large dams, such as the Hoover Dam, and the Three Gorges Dam are intended to last "forever", a period that is not quantified.

World Records

The world record for the largest concrete pour in a single project is the Three Gorges Dam in Hubei Province, China by the Three Gorges Corporation. The amount of concrete used in the construction of the dam is estimated at 16 million cubic metres over 17 years. The previous record was 12.3 million cubic metres held by Itaipu hydropower station in Brazil.

The world record for concrete pumping was set on 7 August 2009 during the construction of the Parbati Hydroelectric Project, near the village of Suind, Himachal Pradesh, India, when the concrete mix was pumped through a vertical height of 715 m (2,346 ft).

The world record for the largest continuously poured concrete raft was achieved in August 2007 in Abu Dhabi by contracting firm Al Habtoor-CCC Joint Venture and the concrete supplier is Unibeton Ready Mix. The pour (a part of the foundation for the Abu Dhabi's Landmark Tower) was 16,000 cubic metres of concrete poured within a two-day period.

The previous record, 13,200 cubic metres poured in 54 hours despite a severe tropical storm requiring the site to be covered with tarpaulins to allow work to continue, was achieved in 1992 by joint Japanese and South Korean consortiums Hazama Corporation and the Samsung C&T Corporation for the construction of the Petronas Towers in Kuala Lumpur, Malaysia.

The world record for largest continuously poured concrete floor was completed 8 November 1997, in Louisville, Kentucky by design-build firm EXXCEL Project Management. The monolithic placement consisted of 225,000 square feet (20,900 m^2) of concrete placed within a 30-hour period, finished to a flatness tolerance of F_F 54.60 and a levelness tolerance of F_L 43.83. This surpassed the previous record by 50% in total volume and 7.5% in total area.

The record for the largest continuously placed underwater concrete pour was completed 18 October 2010, in New Orleans, Louisiana by contractor C. J. Mahan Construction Company, LLC of Grove City, Ohio. The placement consisted of 10,251 cubic yards of concrete placed in a 58.5 hour period using two concrete pumps and two dedicated concrete batch plants. Upon curing, this placement allows the 50,180-square-foot (4,662 m^2) cofferdam to be dewatered approximately 26 feet (7.9 m) below sea level to allow the construction of the IHNC GIWW Sill & Monolith Project to be completed in the dry.

Bridges

More bridges are built using concrete than any other material worldwide, demonstrating continued confidence in the material's performance and durability. Concrete bridges worldwide have a clear track record of flexibility and versatility in terms both of final forms and methods of construction that is hard to match.

Durability, aesthetics, economic solutions, simplified construction and rapid deployment techniques all contribute to making concrete the best construction material for any bridge project, whatever the size, form or intended use. Greater construction flexibility can be realised through the many forms of concrete easily available nationwide, making concrete an adaptable resource suitable for deployment for even the most challenging of bridge types or construction sites.

Benefits

Function and Elegance: The range of structural forms achievable with concrete is limited only by designers' imagination, enabling the creation of elegant bridge structures that blend seamlessly with natural surroundings. Concrete unites both function and elegance in safe, robust structures, regardless of scale. Architectural surface finishes can add to visual impacts while at the same time eliminating the need for cladding, painting and ongoing maintenance activities.

Durability

Concrete is ideally suited to exposure to aggressive environmental conditions, meaning achievable service lives of 120 years or more. Well designed and constructed concrete bridges require only minimum maintenance to keep them in good working condition.

Versatility

Concrete's versatility enables the cost-effective delivery of a wide range of structural forms and associated spans. For instance, solutions including concrete arches, slab decks, beam bridges, box girder, cable stayed, stressed ribbon and inclined frames offer spans ranging from 5 to in excess of 400m.

Speed and Buildability

Adequate pre-planning, precasting of elements and the use of appropriate technology in design and construction can make concrete the cheapest and fastest material for constructing durable, quality bridges. Techniques such as sliding, launching, jacking or modular construction make concrete bridge solutions quick and competitive.

Sustainability

Concrete as a locally sourced material with low environmental and social impact, it lends itself to structures with a typical design life of at least 100 years with low maintenance requirements thereby providing a sustainable solution for bridge construction. Furthermore, up to 95 per cent of concrete and reinforcement used in bridges is recyclable at the end of its viable use.

Economy

Competitive initial construction costs, coupled with reduced inspection and maintenance, means concrete's cost efficiency is very attractive in the long-term.

Seismic Design of Steel Bridges

The Steel Bridge is a through truss, double-deck vertical-lift bridge across the Willamette River in Portland, Oregon, United States, opened in 1912. Its lower deck carries railroad and bicycle/pedestrian traffic, while the upper deck carries road traffic (on the Pacific Highway West No. 1W, former Oregon Route 99W) and light rail (MAX), making the bridge one of the most multimodal in the world. It is the only double-deck bridge with independent lifts in the world and the second oldest vertical-lift bridge in North America, after the nearby Hawthorne Bridge. The bridge links the Rose Quarter and Lloyd District in the east to Old Town Chinatown neighbourhood in the west.

The bridge was completed in 1912 and replaced the Steel Bridge that was built in 1888 as a double-deck swing-span bridge. The 1888 structure was the first railroad bridge across the Willamette River in Portland. Its name originated because steel, instead of wrought iron, was used in its construction, very unusual for the time. When the current Steel Bridge opened, it was simply given its predecessor's name.

The bridge was designed by the engineering firm of Waddell & Harrington, which was based in Kansas City, Missouri, but also had an office in Portland. The structure was built by Union Pacific Railroad and the Oregon-Washington Railroad and Navigation Company (OWR&N) at a cost of $1.7 million (equivalent to $41 million in 2013). It opened in July 1912 to rail traffic and on August 9, 1912, to automobiles.

The 1888 Steel Bridge (upper deck) had been crossed by horse-drawn streetcars from the time of its opening and then by the city's first electric streetcar line starting in November 1889; when the present Steel Bridge opened in 1912, the streetcar lines (all electric by then) moved to it, starting on September 8, 1912. Streetcar service across the Steel continued until August 1, 1948, when the last car lines using it, the Alberta and Broadway Lines, were abandoned. A single line of Portland's once-extensive trolley bus system also used the bridge; the Williams Avenue line crossed the Steel Bridge from February 1937 until October 9, 1949. Many years later, in 1986, electric transit vehicles returned to the bridge in the form of MAX Light Rail and later the Portland Vintage Trolley. In 1950, the Steel Bridge became an important part of a new U.S. 99W highway between Harbor Drive and Interstate Avenue. Harbor Drive was removed in 1974 and replaced with Tom McCall Waterfront Park.

***Figure:** A MAX light rail train crossing the bridge in 2009. TriMet now has four MAX lines, and all of them cross the Steel Bridge. More than 600 MAX trips cross the bridge each weekday.*

In the mid-1980s, the bridge underwent a $10 million rehabilitation, including construction of the MAX light rail line of TriMet. The span was closed to all traffic for two years, starting in June 1984. It reopened on May 31, 1986. Completion and testing of the light-rail tracks and overhead wires across the bridge took place during the months that followed, and the light rail line opened for service on September 5, 1986.

A single-lane viaduct that connected the bridge's east approach to another viaduct (still in existence) that takes traffic from southbound Interstate 5 to Interstate 84 was closed in 1988 and removed in 1989, as part of roadway changes intended to improve traffic flow around the Oregon Convention Centre. The centre was under construction at that time and opened in 1990.

The lower deck of the bridge was threatened by major floods in 1948, 1964, and 1996.

In 2001, a 220-foot (67 m) long and 8-foot (2.4 m) wide cantilevered walkway was installed on the southern side of the bridge's lower deck as part of the Eastbank Esplanade construction, raising to three the number of publicly accessible walkways across the bridge, including the two narrow sidewalks on the upper deck. The bridge is owned by Union Pacific with the upper deck leased to Oregon Department of Transportation, and subleased to TriMet, while the City of Portland is responsible for the approaches.

***Figure:** An Amtrak Talgo train crossing the bridge*

The average daily traffic in 2000 was 23,100 vehicles (including many TriMet buses), 200 MAX trains, 40 freight and Amtrak trains, and 500 bicycles. The construction of the lower-deck walkway connected to the Eastbank Esplanade resulted in a sharp increase in bicycle traffic, with over 2,100 daily bicycle crossings in 2005. MAX traffic across the Steel has tripled since 2000, when only the Gresham–Hillsboro line (now the "Blue Line") was using the bridge, to 605 daily crossings (weekdays) as of 2012. This resulted from the addition of three more MAX lines during that period: the Red, Yellow, Green Lines.

In summer 2008, the upper deck was closed for three weeks to allow a junction to be built at the west end for the MAX Green Line. A change made at that time is that the two inner lanes became restricted to MAX trains only, with cars, buses and other motorized traffic permitted only in the two outer lanes.

In 2012, the Steel Bridge celebrated its 100th birthday. According to *The Oregonian*, it is the hardest working bridge on the Willamette River: "Cars, trucks, freight trains, buses, Amtrak, MAX, pedestrians, bicycles — you carry it all."

Structure and Lift Operation

Figure: *View from roadway during a lift-span opening*

The lift span of the bridge is 211 feet (64 m) long. At low river levels the lower deck is 26 feet (7.9 m) above the water, and 163 feet

(50 m) of vertical clearance is provided when both decks are raised. Because of the independent lifts, the lower deck can be raised to 72 feet (22 m), telescoping into the upper deck but not disturbing it. Each deck has it own counterweights, two for the upper and eight for the lower, totaling 9,000,000 lb (4,100 metric tons).

The machinery house sits atop the upper-deck lift truss. The operator's room is suspended from the top of the lift-span truss, directly below the machinery house, so that the operator can view river traffic as well as the upper deck. After the 2001 addition of a pedestrian walkway on the lower deck, cameras and closed-circuit television monitors were added to allow the operator to view the lower-deck walkway.

Until the bridge's mid-1980s renovation, the crossing gates blocking the roadway and sidewalks during raising of the upper-deck lift span were manually operated, rotated horizontally across the roadway by two "gate tenders", one on each side of the lift span. Small shacks for the gatekeepers were positioned on the roadway deck, between the inner and outer traffic lanes, but they were removed during the 1980s rebuilding and replaced by a new gate tender house positioned above the roadway, in the west lift tower. Powered crossing gates replaced the manual ones, and operation of the gates is now automated, controlled by the bridge operator.

Truss bridge

A truss bridge is a bridge whose load-bearing superstructure is composed of a truss. This truss is a structure of connected elements forming triangular units. The connected elements (typically straight) may be stressed from tension, compression, or sometimes both in response to dynamic loads. Truss bridges are one of the oldest types of modern bridges. The basic types of truss bridges shown in this article have simple designs which could be easily analyzed by nineteenth and early twentieth century engineers. A truss bridge is economical to construct owing to its efficient use of materials.

Design

The nature of a truss allows the analysis of the structure using a few assumptions and the application of Newton's laws of motion according to the branch of physics known as statics. For purposes of analysis, trusses are assumed to be pin jointed where the straight components meet. This assumption means that members of the truss (chords, verticals and diagonals) will act only in tension or compression. A more complex analysis is required where rigid joints impose significant bending loads upon the elements, as in a Vierendeel truss.

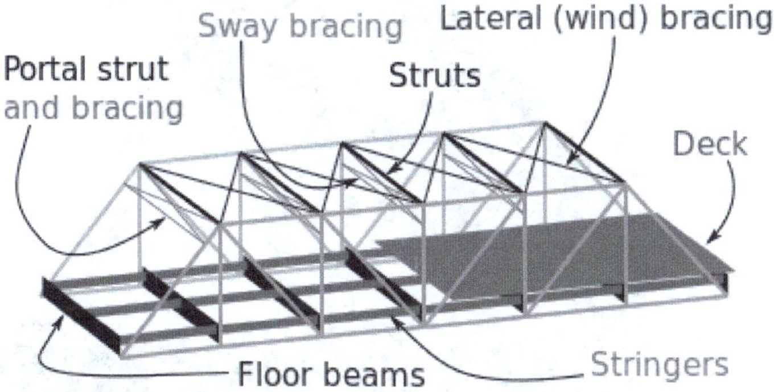

Figure: *The integral members of a truss bridge*

In the bridge illustrated in the infobox at the top, vertical members are in tension, lower horizontal members in tension, shear, and bending, outer diagonal and top members are in compression, while the inner diagonals are in tension.

The central vertical member stabilizes the upper compression member, preventing it from buckling. If the top member is sufficiently stiff then this vertical element may be eliminated. If the lower chord (a horizontal member of a truss) is sufficiently resistant to bending and shear, the outer vertical elements may be eliminated, but with additional strength added to other members in compensation. The ability to distribute the forces in various ways has led to a large variety of truss bridge types. Some types may be more advantageous when wood is employed for compression elements while other types may be easier to erect in particular site conditions, or when the balance between labour, machinery and material costs have certain favourable proportions.

The inclusion of the elements shown is largely an engineering decision based upon economics, being a balance between the costs of raw materials, off-site fabrication, component transportation, on-site erection, the availability of machinery and the cost of labour. In other cases the appearance of the structure may take on greater importance and so influence the design decisions beyond mere matters of economics. Modern materials such as prestressed concrete and fabrication methods, such as automated welding, and the changing price of steel relative to that of labour have significantly influenced the design of modern bridges.

History in the United States

Figure: *Warren-type through-truss of the former Seaboard Air Line Railway, located near the village of Willow, Florida; abandoned since the mid-1980s*

Because wood was so abundant, early truss bridges would typically use carefully fitted timbers for members taking compression and iron rods for tension members, usually constructed as a covered bridge to protect the structure. In 1820 a simple form of truss, Town's lattice truss, was patented, and had the advantage of not requiring high labour skills nor much metal. Few iron truss bridges were built in the United States before 1850.

Truss bridges became a common type of bridge built from the 1870s through the 1930s. Examples of these bridges still remain across the United States, but their numbers are dropping rapidly, as they are demolished and replaced with new structures. As metal slowly started to replace timber, wrought iron bridges in the U.S. started being built on a large scale in the 1870s. Bowstring truss bridges were a common truss design seen during this time, with their arched top chords. Companies like the Massillon Bridge Company of Massillon, Ohio and the King Bridge Company of Cleveland, Ohio became well-known companies, as they marketed their designs to different cities and townships. The bowstring truss design (photo) fell out of favour due to a lack of durability, and gave way to the Pratt truss design, which was stronger. Again, the bridge companies marketed their designs, with the Wrought Iron Bridge Company in the lead. As the 1880s and 1890s progressed, steel began to replace wrought iron as the preferred material. Other truss designs were used during this time, including the camel-back. By the 1910s, many states developed standard plan truss bridges, including steel Warren pony truss bridges.

As the 1920s and 1930s progressed, some states, such as Pennsylvania, continued to build steel truss bridges, including massive steel through truss bridges for long spans. Other states, such as Michigan, used standard plan concrete girder and beam bridges, and only a limited number of truss bridges were built.

Roadbed Types

The truss may carry its roadbed on top, in the middle, or at the bottom of the truss. Bridges with the roadbed at the top or the bottom are the most common as this allows both the top and bottom to be stiffened, forming a box truss. When the roadbed is atop the truss it is called a deck truss (an example of this was the I-35W Mississippi River bridge), when the truss members are both above and below the roadbed, it is called a through truss (an example of this application is the Pulaski Skyway), and where the sides extend above the roadbed but are not connected, a pony truss or half-through truss.

Figure: Deck truss *railroad bridge over the Erie Canal in Lockport, New York*

Sometimes both the upper and lower chords support roadbeds, forming a *double-decked truss.* This can be used to separate rail from road traffic or to separate the two directions of automobile traffic and so avoiding the likelihood of head-on collisions.

Since through truss bridges have supports located over the bridge deck, they are susceptible to being hit by overheight loads when used on highways - the I-5 Skagit River bridge collapsed after such a strike, and such impacts were common and required frequent repairs before the collapse.

Figure: *The four span* through truss *General Hertzog Bridge over the Orange River at Aliwal North carries vehicular traffic.*

Truss Types Used in Bridges

Bridges are, many times, the best examples of truss use to the common person. There are many types of designs, many dating back hundreds of years. Below are some of the more common types and designs.

Allan Truss

The Allan Truss, designed by Percy Allan, is partly based on the Howe truss. The first Allan truss was completed on 13 August 1894 over Glennies Creek at Camberwell, New South Wales and the last Allan truss bridge was built over Mill Creek near Wisemans Ferry in 1929.

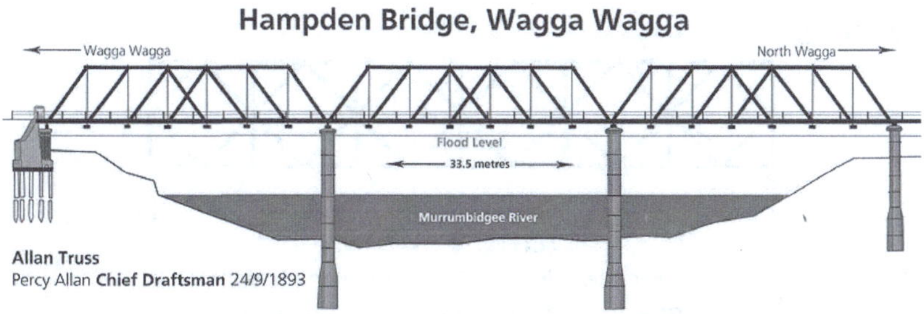

Figure: *Allan Truss illustrated*

Figure: *Hampden Bridge showing the Allan truss design*

Completed in March 1895, the Tharwa Bridge located at Tharwa, Australian Capital Territory, was the second Allan truss to be built, the oldest surviving bridge in the Australian Capital Territory and the oldest, longest in continuous use Allan truss bridge.

Completed in November 1895, the Hampden Bridge in Wagga Wagga, New South Wales, Australia, the first of the Allan truss bridges with overhead bracing, was originally designed as a steel bridge but was constructed with timber to reduce cost. In his design, Allan used Australian ironbark for its strength.

A similar bridge also designed by Percy Allen is the Victoria Bridge on Prince Street Picton, New South Wales. Also constructed of ironbark, the bridge is still in use today for pedestrian and light traffic.

Bailey Bridge

Figure: *Bailey truss illustrated.*

Figure: *Bailey bridge over the Meurthe River, France.*

Designed for military use, the prefabricated and standardized truss elements may be easily combined in various configurations to adapt to the needs at the site. In the image at right, note the use of doubled prefabrications to adapt to the span and load requirements. In other applications the trusses may be stacked vertically.

Baltimore Truss

Figure: *Baltimore truss illustrated.*

The Baltimore truss is a subclass of the Pratt truss. A Baltimore truss has additional bracing in the lower section of the truss to prevent buckling in the compression members and to control deflection. It is mainly used for train bridges, boasting a simple and very strong design.

Bollman Truss

Figure: *Bollman truss in Savage, Maryland. Built in 1869, moved to Savage in 1887. It is still in use today as a pedestrian bridge. 39°82 5.423 N 76°492 30.333 W*

The Bollman Truss Railroad Bridge at Savage, Maryland is the only surviving example of a revolutionary design in the history of American bridge engineering. The type was named for its inventor, Wendel Bollman, a self-educated Baltimore engineer. It was the first successful all-metal bridge design (patented in 1852) to be adopted and consistently used on a railroad. The design employs wrought iron tension members and cast iron compression members. The use of multiple independent tension elements reduces the likelihood of catastrophic failure. The structure was also easy to assemble.

The Wells Creek Bollman Bridge is the only other bridge designed by Wendel Bollman still in existence, but it is a Warren truss configuration.

Bowstring Arch Truss

Figure: *Bowstring truss illustrated*

Figure: *A bowstring truss bridge, in London, Ontario, Canada*

The bowstring arch through truss bridge was patented in 1841 by Squire Whipple. While similar in appearance to a Tied-arch bridge, a bowstring truss is a truss and thus has diagonal load-bearing members. These diagonals result in a structure that more closely matches a Parker truss or Pratt truss than a true arch.

Brown Truss

This type of truss is particularly suited for timber structures that use iron rods as tension members.

The Brown truss is a box truss that is a through truss (as contrasted with a deck truss) and consists of diagonal cross compression members connected to horizontal top and bottom stringers. There may be vertical or almost vertical tension members (the diagram shows these members, while the patent application diagram does not) but there are no vertical members in compression. In practice, when used in a covered bridge, the most common application, the truss is protected with outside sheathing.

The floor and roof are also trusses, but are horizontal and serve to give the truss rigidity. The bottoms of the diagonals tend to protrude below the sheathing. The Brown truss is noted for economy of materials as it can be built with very little metal.

Burr Arch Truss

Figure: *A covered bridge with a Burr Arch Truss structure*

This combines an arch with a truss to form a structure both strong and rigid.

Cantilevered Truss

Figure: *Forth Bridge*

Most trusses have the lower chord under tension and the upper chord under compression. In a cantilever truss the situation is reversed, at least over a portion of the span.

The typical cantilever truss bridge is a *balanced cantilever*, which enables the construction to proceed outward from a central vertical spar in each direction. Usually these are built in pairs until the outer sections may be anchored to footings. A central gap, if present, can then be filled by lifting a conventional truss into place or by building it in place using a *travelling support*.

Fink Truss

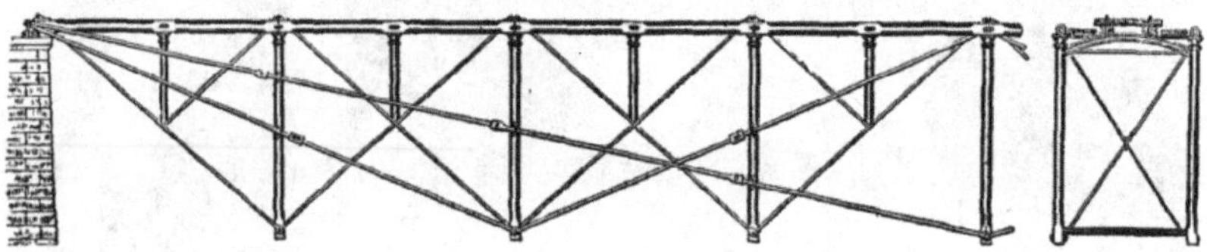

Figure: *Fink Truss (half span and cross section)*

The Fink truss was designed by Albert Fink of Germany in the 1860s. This type of bridge was popular with the Baltimore and Ohio Railroad. The Appomattox High Bridge on the Norfolk and Western Railroad included 21 Fink deck truss spans from 1869 until their replacement in 1886.

Howe Truss

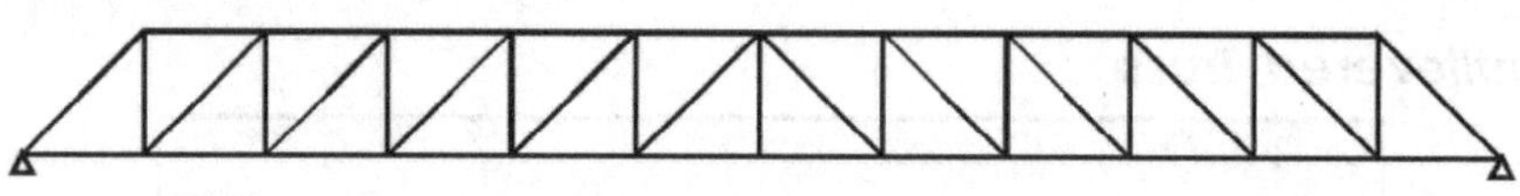

Figure: *Howe truss illustrated - the diagonals are under compression under balanced loading*

The relatively rare Howe truss, patented in 1840 by Massachusetts millwright William Howe, includes vertical members and diagonals that slope up towards the centre, the opposite of the Pratt truss. In contrast to the Pratt Truss, the diagonal web members are in compression and the vertical web members are in tension. Examples include Jay Bridge in Jay, New York, and Sandy Creek Covered Bridge in Jefferson County, Missouri.

K Truss

A truss in the form of a K due to the orientation of the vertical member and two oblique members in each panel. An example is the Südbrücke rail bridge over the River Rhine, Mainz, Germany.

Kingpost Truss

***Figure:** King Post Truss*

One of the simplest truss styles to implement, the king post consists of two angled supports leaning into a common vertical support.

Lattice Truss (Town's Lattice Truss)

***Figure:** Lattice, or Warren quadrangular truss illustrated*

***Figure:** Plank lattice truss of a covered bridge*

This type of bridge uses a substantial number of lightweight elements, easing the task of construction. Truss elements are usually of wood, iron, or steel.

Lenticular Truss

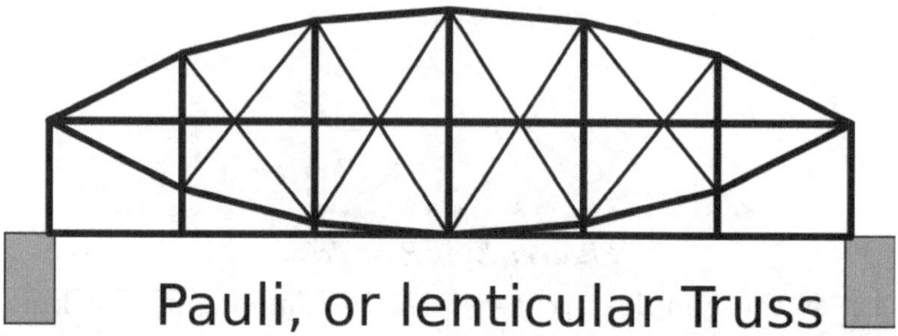

***Figure:** Lenticular, or Pauli truss illustrated*

***Figure:** Royal Albert Bridge under construction, 1859*

A lenticular truss bridge includes a lens-shape truss, with trusses between an upper arch that curves up and then down to end points, and a lower arch that curves down and then up to meet at the same end points. Where the arches extend above and below the roadbed, it is a lenticular pony truss bridge.

One type of lenticular truss consists of arcuate upper compression chords and lower eyebar chain tension links. The Royal Albert Bridge (United Kingdom) uses a single tubular upper chord. As the horizontal tension and compression forces are balanced these horizontal forces are not transferred to the supporting pylons (as is the case with most arch types). This in turn enables the truss to be fabricated on the ground and then to be raised by jacking as supporting masonry pylons are constructed. This truss has been used in the construction of a stadium, with the upper chords of parallel trusses supporting a roof that may be rolled back. The Smithfield Street Bridge in Pittsburgh, Pennsylvania is another example of this type.

An example of a lenticular pony truss bridge that uses regular spans of iron is the Turn-of-River Bridge designed and manufactured by the Berlin Iron Bridge Co..

Long Truss

Designed by Stephen H. Long in 1830. The design resembles a Howe truss, but is entirely made of wood instead of a combination of wood and metal. The longest surviving example is the Eldean Covered Bridge north of Troy, Ohio, spanning 224 feet. One of the earliest examples is the Old Blenheim Bridge, which with a span of 210 feet and a total length of 232 feet long was the second-longest covered bridge in the United States, until its destruction from flooding in 2011.

The Busching bridge, unfortunately used as an example of a Long truss, is an example of a Howe truss, as the verticals are metal rods. A Long truss has timber verticals.

Parker (Camelback) Truss

"Camelback bridge" redirects here.

Figure: *Parker truss illustrated.*

A Parker truss bridge is a Pratt truss design with a polygonal upper chord. A "camelback" is a subset of the Parker type, where the upper chord consists of exactly five segments. An example of a Parker truss is the Traffic Bridge in Saskatoon, Canada. An example of a camelback truss is the Woolsey Bridge near Woolsey, Arkansas.

Figure: *The Woolsey Bridge is an example of a Parker camelback truss*

Pegram Truss

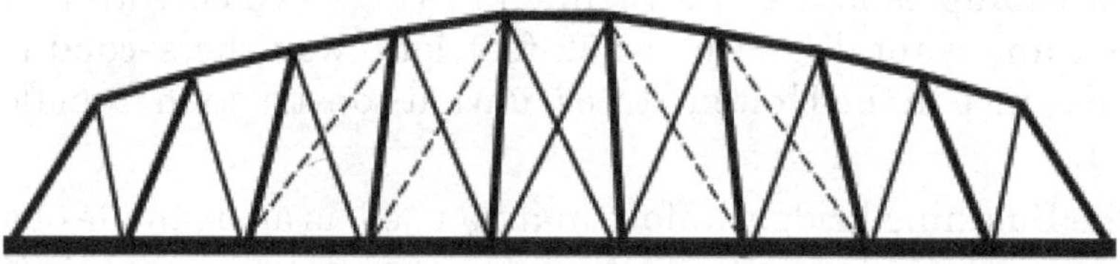

Figure: *Pegram Truss*

The Pegram truss is a hybrid between the Warren and Parker trusses where the upper chords are all of equal length and the lower chords are longer than the corresponding upper chord. Because of the difference in upper and lower chord length, each panel is not square. The members which would be vertical in a Parker truss vary from near vertical in the centre of the span to diagonal near each end (like a Warren truss). George H. Pegram, while the chief engineer of Edge Moor Iron Company in Wilmington, Delaware, patented this truss design in 1885.

The Pegram truss consists of a Parker type design with the vertical posts leaning towards the centre at an angle between 60 and 75°. The variable post angle and constant chord length allowed steel in existing bridges to be recycled into a new span using the Pegram truss design. This design also facilitated reassembly and permitted a bridge to be adjusted to fit different span lengths. There are twelve known remaining Pegram span bridges in the United States with seven in Idaho, two in Kansas, and one each in California, Washington, and Utah.

Pennsylvania (Petit) Truss

The Pennsylvania (Petit) truss is a variation on the Pratt truss. The Pratt truss includes braced diagonal members in all panels; the Pennsylvania truss adds to this design half-length struts or ties in the top, bottom, or both parts of the panels.

It is named after the Pennsylvania Railroad, which pioneered this design. It was once used for hundreds of bridges in the United States, but fell out of favour in the 1930s, and very few bridges of this design remain. Examples of this truss type include the Schell Bridge in Northfield, Massachusetts, the Inclined Plane Bridge in Johnstown, Pennsylvania, and the Healdsburg Memorial Bridge in Healdsburg, California.

Post Truss

A Post truss is a hybrid between a Warren truss and a double-intersection Pratt truss. Invented in 1863 by Simeon S. Post, it is occasionally referred to as a *Post patent truss* although he never received a patent for it. The Ponakin Bridge and the Bell Ford Bridge are two examples of this truss.

Pratt Truss

A Pratt truss includes vertical members and diagonals that slope down towards the centre, the opposite of the Howe truss. The interior diagonals are under tension under balanced loading and vertical elements under compression. If pure tension elements are used in the diagonals (such as eyebars) then crossing elements may be needed near the centre to accept concentrated live loads as they traverse the span. It can be subdivided, creating Y- and K-shaped patterns. The Pratt Truss was invented in 1844 by Thomas and Caleb Pratt. This truss is practical for use with spans up to 250 feet and was a common configuration for railroad bridges as truss bridges moved from wood to metal. They are statically determinate bridges, which lend themselves well to long spans. They were common in the United States between 1844 and the early 20th century.

Examples of a Pratt Truss bridge are the Governor's Bridge in Maryland, Dearborn River High Bridge near Augusta, Montana, built in 1897, and the Fair Oaks Bridge in Fair Oaks, California, built 1907-1909.

Queenpost Truss

The queenpost truss, sometimes *queen post* or *queenspost*, is similar to a king post truss in that the outer supports are angled towards the centre of the structure. The primary difference is the horizontal extension at the centre which relies on beam action to provide mechanical stability. This truss style is only suitable for relatively short spans.

Thatcher Truss

The Thatcher truss combines some of the characteristics of a Pratt truss with diagonals under tension and a Howe truss with diagonals under compression. It is quite rare.

Truss Arch

A truss arch may contain all horizontal forces within the arch itself, or alternatively may be either a thrust arch consisting of a truss, or of two arcuate sections pinned at the apex. The latter form is common when the bridge is constructed as cantilever segments from each side as in the Navajo Bridge.

Waddell Truss

Patented 1894 (U.S. Patent 529,220) its simplicity eases erection at the site. It was intended to be used as a railroad bridge.

Warren Truss

The Warren truss was patented in 1848 by its designers James Warren and Willoughby Theobald Monzani, and consists of longitudinal members joined only by angled cross-members, forming alternately inverted equilateral triangle-shaped spaces along its length, ensuring that no individual strut, beam, or tie is subject to bending or torsional straining forces, but only to tension or compression. Loads on the diagonals alternate between compression and tension (approaching the centre), with no vertical elements, while elements near the centre must support both tension and compression in response to live loads. This configuration combines strength with economy of materials and can therefore be relatively light. The girders being of equal length, it is ideal for use in prefabricated modular bridges. It is an improvement over the Neville truss which uses a spacing configuration of isosceles triangles.

Warren truss construction has also been used in airframe construction for aircraft since the 1920s, mostly for smaller aircraft fuselages, using chrome molybdenum alloy steel tubing, with popular aircraft such as the Piper J-3 *Cub*. One of the earliest uses for the Warren truss design in aircraft design was for the interplane wing strut layout, as seen in a nose-on view, on the Italian World War I Ansaldo SVA series of fast reconnaissance biplanes, which were among the fastest aircraft of the First World War era. Warren truss construction is still used today for some homebuilt aircraft fuselage designs, that essentially use the same 1920s-era design philosophies in the 21st century.

Whipple Truss

A Whipple truss is usually considered a subclass of the Pratt truss because the diagonal members are designed to work in tension. The main characteristic of a whipple truss is that the tension members are elongated, usually thin, at a shallow angle and cross two or more bays.

Vierendeel Truss

The Vierendeel truss, unlike common pin-jointed trusses, imposes significant bending forces upon its members — but this in turn allows the elimination of many diagonal elements. It is a structure where the members are not triangulated but form rectangular openings, and is a frame with fixed joints that are capable of transferring and resisting bending moments. While rare as a bridge type due to higher costs compared to a triangulated truss, it is commonly employed in modern building construction as it allows the resolution of gross shear forces against the frame elements while retaining rectangular openings between columns. This is advantageous both in allowing flexibility in the use of the building space and freedom in selection of the building's outer curtain wall, which affects both interior and exterior styling aspects.

4

Cable-stayed Bridge

A cable-stayed bridge has one or more *towers* (or *pylons*), from which cables support the bridge deck.

There are two major classes of cable-stayed bridges: *harp* and *fan*.

In the *harp* design, the cables are nearly parallel so that the height of their attachment to the tower is proportional to the distance from the tower to their mounting on the deck.

In the *fan* design, the cables all connect to or pass over the top of the towers. The fan design is structurally superior with minimum moment applied to the towers but for practical reasons the modified fan is preferred especially where many cables are necessary. In the modified fan arrangement the cables terminate near to the top of the tower but are spaced from each other sufficiently to allow better termination, improved environmental protection, and good access to individual cables for maintenance.

The cable-stayed bridge is optimal for spans longer than cantilever bridges, and shorter than suspension bridges. This is the range where cantilever bridges would rapidly grow heavier if the span were lengthened, and suspension bridge cabling would not be more economical if the span were shortened.

History

Cable-stayed bridges date back to 1595, where designs were found in *Machinae Novae*, a book by Venetian inventor Fausto Veranzio. Many early suspension bridges were cable-stayed construction, including the 1817 footbridge Dryburgh Bridge, James Dredge's patented Victoria Bridge, Bath (1836), and the later Albert Bridge

(1872) and Brooklyn Bridge (1883). Their designers found that the combination of technologies created a stiffer bridge. John A. Roebling took particular advantage of this to limit deformations due to railway loads in the Niagara Falls Suspension Bridge.

The earliest known surviving example of a true cable-stayed bridge in the United States is E.E. Runyon's largely intact steel or iron bridge with wooden stringers and decking in Bluff Dale, Texas (1890), or his weeks-earlier but ruined Barton Creek Bridge between Huckabay, Texas and Gordon, Texas (1889 or 1890). In the twentieth century, early examples of cable-stayed bridges included A. Gisclard's unusual Cassagnes bridge (1899), in which the horizontal part of the cable forces is balanced by a separate horizontal tie cable, preventing significant compression in the deck, and G. Leinekugel le Coq's bridge at Lézardrieux in Brittany (1924). Eduardo Torroja designed a cable-stayed aqueduct at Tempul in 1926. Albert Caquot's 1952 concrete-decked cable-stayed bridge over the Donzère-Mondragon canal at Pierrelatte is one of the first of the modern type, but had little influence on later development. The steel-decked Strömsund Bridge designed by Franz Dischinger (1955) is therefore more often cited as the first modern cable-stayed bridge.

Other key pioneers included Fabrizio de Miranda, Riccardo Morandi and Fritz Leonhardt. Early bridges from this period used very few stay cables, as in the Theodor Heuss Bridge (1958). However, this involves substantial erection costs, and more modern structures tend to use many more cables to ensure greater economy.

Comparison with Suspension Bridge

Cable-stayed bridges may appear to be similar to suspension bridges, but in fact they are quite different in principle and in their construction.

In suspension bridges, large main cables (normally 2) hang between the towers (normally 2), and are anchored at each end to the ground. This can be difficult to implement when ground conditions are poor. The main cables, which are free to move on bearings in the towers, bear the load of the bridge deck. Before the deck is installed, the cables are under tension from their own weight. Along the main cables smaller cables or rods connect to the bridge deck, which is lifted in sections. As this is done, the tension in the cables increases, as it does with the live load of traffic crossing the bridge. The tension on the main cables is transferred to the ground at the anchorages and by downwards tug on the towers.

Difference Between Types of Bridges

In the cable-stayed bridge, the towers are the primary load-bearing structures which transmit the bridge loads to the ground. A cantilever approach is often used to support the bridge deck near the towers, but lengths further from them are supported by cables running directly to the towers.

This has the disadvantage, compared to the suspension bridge, that the cables pull to the sides as opposed to directly up, requiring the bridge deck to be stronger to resist the resulting horizontal compression loads; but has the advantage of not requiring firm anchorages to resist the horizontal pull of the main cables of the suspension bridge. By design all static horizontal forces of the cable-stayed bridge are balanced so that the supporting towers do not tend to tilt or slide, needing only to resist horizontal forces from the live loads.

Key advantages of the cable-stayed form are as follows:

- much greater stiffness than the suspension bridge, so that deformations of the deck under live loads are reduced
- can be constructed by cantilevering out from the tower - the cables act both as temporary and permanent supports to the bridge deck
- for a symmetrical bridge (i.e. spans on either side of the tower are the same), the horizontal forces balance and large ground anchorages are not required

Cantilever-spar Cable-stayed Bridge

Far more radical in its structure, the Puente del Alamillo (1992) uses a single cantilever spar on one side of the span, with cables on one side only to support the bridge deck. Unlike other cable-stayed types, this bridge exerts considerable overturning force upon its foundation and the spar must resist the bending caused by the cables, as the cable forces are not balanced by opposing cables. The spar of this particular bridge forms the gnomon of a large garden sundial. Related bridges by the architect Santiago Calatrava include the Puente de la Mujer (2001), Sundial Bridge (2004) and Chords Bridge (2008).

Multiple-span Cable-stayed Bridge

Cable-stayed bridges with more than three spans involve significantly more challenging designs than do 2-span or 3-span structures.

In a 2-span or 3-span cable-stayed bridge, the loads from the main spans are normally anchored back near the end abutments by stays in the end spans. For more spans, this is not the case and the bridge structure is less stiff overall. This can create difficulties in both the design of the deck and the pylons.

Examples of multiple-span structures in which this is the case include Ting Kau Bridge, where additional 'cross-bracing' stays are used to stabilise the pylons; Millau Viaduct and Mezcala Bridge, where twin-legged towers are used; and General Rafael Urdaneta Bridge, where very stiff multi-legged frame towers were adopted. A similar situation with a suspension bridge is found at both the Great Seto Bridge and San Francisco – Oakland Bay Bridge where additional anchorage piers are required after every set of three suspension spans - this solution can also be adapted for cable-stayed bridges.

Extradosed Bridge

The extradosed bridge is a cable-stayed bridge but with a more substantial bridge deck that, being stiffer and stronger, allows the cables to be omitted close to the tower and for the towers to be lower in proportion to the span. The first extradosed bridge in the US is the Pearl Harbor Memorial Bridge (Connecticut) in New Haven, Connecticut, which is scheduled to fully open in 2016. A new extradosed bridge is also being planned to cross the Saint Croix River between Bayport, Minnesota and Houlton, Wisconsin in the Twin Cities.

Cable-stayed Cradle-system Bridge

A cradle system carries the strands within the stays from bridge deck to bridge deck, as a continuous element, eliminating anchorages in the pylons. Each epoxy-coated steel strand is carried inside the cradle in a one-inch (2.54 cm) steel tube. Each strand acts independently, allowing for removal, inspection and replacement of individual strands. The first two such bridges are the Penobscot Narrows Bridge, completed in 2006, and the Veterans' Glass City Skyway, completed in 2007.

Related Bridge Types

Self Anchored Suspension Bridge

A self-anchored suspension bridge has some similarity in principle to the cable-stayed type in that tension forces that prevent the deck from dropping are converted into compression forces vertically in the tower and horizontally along the deck structure. It is also related to the suspension bridge in having arcuate main cables with suspender cables, although the self-anchored type lacks the heavy cable anchorages of the ordinary suspension bridge.

Unlike either a cable stayed bridge or a suspension bridge, the self-anchored suspension bridge must be supported by falsework during construction and so it is more expensive to construct.

Notable Cable-stayed Bridges

- Cable Bridge, the second cable-stayed bridge in the United States, crosses the Columbia River and connects Pasco, Washington to Kennewick, Washington.
- Centennial Bridge, a six-lane vehicular bridge that crosses the Panama Canal with a total length of 1.05 km (3,451 ft).
- Christopher S. Bond Bridge, crosses the Missouri River at Kansas City, opened in September 2010
- Clark Bridge, crosses the Mississippi River at Alton, Illinois; subject of the Nova documentary *Super Bridge.*
- Dames Point Bridge, also known as the *Napoleon Bonaparte Broward Bridge*, it crosses the St. Johns River near Blount Island in Metropolitan Jacksonville, Florida, opened in 1990.
- Deh Cho Bridge, the first major bridge in Canada's Northwest Territories, 1100 metres across the ice-choked Mackenzie River, opened late 2012.
- Denver Millennium Bridge, a 130-foot (40 m) long pedestrian bridge which won the Gold Award in 2003 from the New York Association of Consulting Engineers. The bridge is a focal point of the film "Imagine That" starring Eddie Murphy.
- Erasmus Bridge, crosses the Nieuwe Maas in Rotterdam, Netherlands. The southern span of the bridge has an 89-metre-long (292 ft) bascule bridge for ships that cannot pass under the bridge. The bascule bridge is the largest and heaviest in West Europe and has the largest panel of its type in the world.
- Greenville Bridge, the third longest cable-stayed bridge main span, 1,378-foot (420 m), in the United States, and fourth longest in North America, crossing the Mississippi River between Lake Village, Arkansas and Greenville, Mississippi.

- Jesús Izcoa Moure Bridge, the first Cable-stayed bridge in Puerto Rico and the Caribbean.
- Jiashao Bridge, Zhejiang Province, China. The bridge is an eight lane structure that spans 10,100 metres across Hangzhou Bay, connecting Jiaxing and Shaoxing, two cities of Zhejiang province. It was opened on 23rd July, 2013 and is currently the longest and widest cable-stayed bridge in the world.
- John James Audubon Bridge (Mississippi River): The longest cable-stayed bridge in the Western Hemisphere, crossing the Mississippi River between New Roads, Louisiana and St. Francisville, Louisiana.
- Juscelino Kubitschek bridge, a unique bridge featuring three 200-foot-tall (61 m) asymmetrical steel arches that crisscross diagonally. Decks are suspended by steel cables alternating and interlacing at each side.
- Kap Shui Mun Bridge, carries both six-lane road and dual-track railway traffic, with a main span of 430 metres and an overall length of 1323 metres, connecting Ma Wan and Lantau Island in Hong Kong as part of the Lantau Link serving Hong Kong International Airport.
- Indiano Bridge, crosses the Arno River in Florence, Italy, 1978, is the first earth-anchored cable-stayed steel bridge in the world with central planes of cables, single inclined masts, vented deck.
- Leonard P. Zakim Bunker Hill Memorial Bridge, the world's widest cable-stayed bridge with a 56 m (183 ft) wide bridge deck; carries 10 lanes of Interstate 93 over the Charles River in Boston, Massachusetts. It is also the first cable-stayed bridge with an asymmetrical deck design.
- Luling Bridge, crosses the Mississippi River in Destrehan, Louisiana; it was the first cable-stayed bridge in the United States carrying an interstate highway
- Lyne Bridge, crosses the M25 motorway near Chertsey, England, 1979, is one of the few cable-stayed bridges to carry a heavy railway (the Chertsey Branch Line).
- Megyeri Bridge, crosses the Danube in Budapest, Hungary, opened in 2008
- Millau Viaduct, the bridge with the tallest piers in the world: 341 metres tall (1,118 ft) and roadway 266 metres high (886 ft),

spanning the Tarn River in France. With a total length of 2 460 m and seven towers, it also has the longest cable-stayed suspended deck in the world.

- Most SNP (Nový most), the world's longest cable-stayed bridge in category with one pylon and with one cable-stayed plane, spanning Danube in Bratislava, Slovakia.The main span is 303 metres, total length 430.8 metres. The only member of World Federation of Great Towers that is primarily used as a bridge. It houses a flying-saucer restaurant at the top of pylon 85 metres tall.
- Natcher Bridge, one of the first cable-stayed bridges on the Ohio River.
- New Railroad Bridge, the first bridge in Europe to use cable-stayed girder system in the railroad industry, connects two banks of the Sava in Belgrade.
- North Avenue Bridge, crosses the Chicago River in Chicago, Illinois, utilizes cable-stayed elements to avoid the full-size abutments required for a pure suspension bridge, while providing additional support with a suspension cable to minimize the height of the bridge's support towers.
- John O'Connell Bridge, crosses the Sitka Channel, between Japonski Island and Baranof Island in Sitka, Alaska, opened in 1972.
- Octavio Frias de Oliveira bridge, crosses the Pinheiros River in São Paulo, 2008. It has a 138 metres (453 ft)-high pylon under which two stayed roads cross each other turning 90° to the opposite bank of the river.
- Oresund Bridge, a combined two-track rail and four-lane road bridge with a main span of 490 metres and a total length of 7.85 km, crossing the Öresund between Malmö, Sweden and the Danish Capital Region.
- Penang Bridge, connects Penang island with the Malay Peninsula, with its longest span measuring 225 m.
- Pont de Normandie, crosses the Seine in Normandy, France - briefly the world's longest cable-stayed bridge.
- Port Mann Bridge, crosses the Fraser River in British Columbia, Canada. It is currently the widest bridge in the world at 65 metres (213 ft) carrying 10 lanes of the Trans Canada Highway.

- Price Bridge (Passerelle George Price) spanning Belgium's Canal du Centre near Mons, commemorates George Lawrence Price, who died at its predecessor in the last minutes of World War I.
- Queen Elizabeth II Bridge, crosses the River Thames in southeast England.
- Rande Bridge in Spain near Vigo is the highway cable-stayed bridge with the longest and slenderest span in the world at the time of construction (1973–1977). Three long spans of 148 m + 400 m + 148 m. Pylons in concrete, girder in steel.
- Rama IX Bridge, Bangkok, 450 m main span, 6 traffic lanes, completed 1987, at that date the second longest centre span in the world.
- Rio-Antirio bridge, crosses the Gulf of Corinth near Patras, Greece. At a total length of 2 880 m and four towers, it has the second longest cable-stayed suspended deck (2 258 m long) in the world.
- Russky Bridge, the cable-stayed bridge with the world's longest span, at 1,104 metres.
- Second Severn Crossing between England and Wales is 3.186 miles (5.127 km) long, consisting of a single central navigation span over the "Shoots" channel and approach viaducts on either side. The Second Severn Crossing Bridge uses Freyssinet cable stays.
- Skybridge, the world's longest transit-only bridge, spanning the Fraser River between New Westminster and Surrey, BC, Canada.
- Stonecutters Bridge, a 1,596 metre-long dual 3-lane high-level cable stayed bridge in Hong Kong, with a clear span of 1,018 metres. The bridge straddles the Rambler Channel at the entrance to the busy Kwai Chung Container port.
- Sundial Bridge, an award-winning single-spar pedestrian bridge spanning the Sacramento River in Redding, California.
- Sunshine Skyway Bridge, the world's longest bridge with a cable-stayed main span; carries I-275 across Tampa Bay south of St. Petersburg, Florida, United States. The very similar looking Oresund bridge is slightly shorter but has a longer main span.

- Suramadu Bridge in Indonesia is the longest cable-stayed bridge in Southeast Asia and Indonesia. It connects the city of Surabaya in Java with the island of Madura. Its main span is 818 metres long.
- Surgut Bridge, the longest single-pylon cable-stayed bridge in the world, crossing the Ob River in Siberia.
- Sutong Bridge, crosses the Yangtze River Delta in eastern China. It has the longest span of any cable-stayed bridge at 1,088 metres (3,570 ft), surpassing Japan's Tatara Bridge for the world's longest cable-stayed main span, but its overall length is shorter than the Sunshine Skyway Bridge in the United States.
- Tatara Bridge, has the second largest span of any cable-stayed bridge at 890 metres (2,920 ft), part of a series of bridges connecting Honshû and Shikoku in Japan.
- Ting Kau Bridge, the world's first major four-span (three towers) cable-stayed bridge, forming part of the road network connecting Hong Kong International Airport to other parts of Hong Kong, China.
- Vasco da Gama Bridge in Lisbon, Portugal is the longest bridge in Europe, with a total length of 17.2 km (10.7 mi), including 0.829 km (0.5 mi) for the main bridge, 11.5 km (7.1 mi) in viaducts, and 4.8 km (3.0 mi) in extension roads.
- Veterans' Glass City Skyway in Toledo, Ohio has a total length of 8,800 feet (2 682 m). The bridge is the most expensive project ever undertaken by the Ohio Department of Transportation (ODOT), costing approximately US$237 million. It carries six lanes of Interstate 280 across the Maumee River.
- Vidyasagar Setu (2nd Hooghly Bridge) in Kolkata, West Bengal, India is the second longest cable-stayed bridge in the Indian subcontinent, with the main span measuring 457.2 m (1,500 ft) and total length being 823 m (2,700 ft).
- Zhivopisny Bridge in Moscow, Russia, is the highest cable-stayed bridge in Europe.
- Envigado bridge in Colombia. This bridge was conceived as representation of the effort and persistence of the community to overcome the convulsed situation that prevailed in the country at the time of construction.

- Xiamen-Zhangzhou Cross Sea Bridge, China. The bridge stretches across the Jiulong River, from Haicang District in Xiamen to Fugong Town in Zhangzhou's Longhai. The bridge has an overall length of 9,330 metres and has a six lane structure. It was opened on 28th of May, 2013 and was the longest cable-stayed-bridge in the world until July 23rd, 2013 when the Jiashao Bridge was opened.

Soil Foundation Structure Interaction

Foundation types depend primarily on the depth and safe bearing pressures of the bearing stratum, also restrictions placed on differential settlement due to the type of bridge deck. Generally in the case of simply supported bridge decks differential settlements of about 20 to 25 mm can be tolerated, whereas multi-span continuous decks 10 mm is usually considered as a maximum.

Bridge foundations generally fall into two categories:

i. Strip footings, one for each pier and abutment. However, it is sometimes convenient to split the deck into two halves longitudinally along the centre line, this is then continued to the footing.

ii. Piled foundations.

It is possible to have a combination of both (i.e. piers being piled with abutments on strip footings).

Design Considerations

The design of foundations comprise of the following stages :

i. From the site investigation report decide upon which stratum to impose the structure load and its safe bearing pressure.

ii. Select the type of foundation, possibly comparing the suitability of several types.

iii. Design the foundation to transfer and distribute the loads from the structure to the ground. Ensure that the factor of safety against shear failure in the soil is not reached and settlement is within the allowable limits.

Strip Footings

The overall size of strip footings is determined by considering the effects of vertical and rotational loads. The combination of these two must neither exceed the safe bearing capacity of the stratum or produce uplift. The thickness of the footings is generally about 0.8 to 1.0 m but must be capable of withstanding moments and shears produced by piers or abutments.

The critical shearing stress may be assumed to occur on a plane at a distance equal to the effective depth of the base from the face of the column. Cover to reinforcement should never be less than values given in BS 5400, and crack control calculation must be carried out to ensure the crack width is less than 0.25mm. Cover to reinforcement will need to be increased to comply with BS 8500 requirements.

Piled Foundations

The type of piles generally used for bridge foundations are :

a. Driven Piles; preformed piles of concrete or steel driven by blows of a power hammer or jacked into the ground.

b. Preformed Driven Cast In-Situ Piles; formed by driving a hollow steel tube with a closed end and filling the tube with concrete.

c. Driven Cast In-Situ Piles; formed by driving a hollow steel tube with a closed end and filling the tube with concrete, simultaneously withdrawing the tube.

d. Bored and Cast In-Situ Piles; formed by boring a hole and filling it with concrete.

e. to c. are known as displacement piles, and the problems of calculating the load carrying capacity and settlement require a different approach to that for bored piles.

Driven type piles can, depending on the strata, be either end bearing or friction piles; sometimes a combination of both. Bored piles are generally end bearing and are often of large diameter. To increase their bearing capacity the bottom can be under-reamed to produce a greater bearing area. However, additional safety precautions are required with larger diameter piles.

A specialist form of pile consisting of stone aggregate consolidated by water or air using the 'Vibroflotation' technique is suitable in some granular soils. Choice of pile type depends largely on the strata which they pass through, none of them however give the most economic and satisfactory solution under all conditions.

The art of selecting the right sort of pile lies in rejecting all those types which are obviously unsuited to the particular set of circumstances and then choosing from those which remain, the one which produces the most economical solution. Concurrently with the choice of pile type must go the choice of the strata which will carry the main loads from the structure, because this very often influences the choice.

In most all cases the rejection of conventional pad or strip foundations arises because the computed settlement is more than the structure can safely withstand and hence the main purpose of the piled foundation will be to reduce this settlement. It follows, therefore, that if more compressible strata exists within reasonable distance of the surface, it is very desirable that a high proportion of the foundation load should be carried by this more stable strata; the ideal solution is where piles support the load wholly in end bearing on hard rock where the settlement will be negligible. It follows that piles wholly embedded in the same soil that would under-lie a conventional foundation has very little effect in reducing settlement. With soft normally consolidated alluvial clays, the remoulding effect of driven piles may well increase the settlement of the soil under its own dead weight and thus increase the settlement of the foundation itself.

Aspects of Design of Piled Foundations Which Influence Choice of Pile Type

All foundations must satisfy two criteria, no shear failure in the soil and no excessive settlement; piled foundations also have to meet this criteria. There are well established methods for ensuring that the first criteria is met, but the second presents more of a problem. The working load of an individual pile is based on providing an adequate factor of safety against the soil under the toe failing in shear and the adhesion between the shaft and the soil surrounding it passing its ultimate value and the whole pile sinking further into the ground. There are basically four methods for assessing this effect :

i. Through soil parameters i.e. summing shaft friction and bearing capacity. The ultimate bearing capacity is usually modified to compensate for the driving effect of the pile.

ii. By means of test piles.

iii. By means of dynamic formulae i.e. Hiley formulae which equates the energy required to drive the pile with its ultimate bearing capacity.

iv. Piling contractors 'know how'.

5

Bridge Components

Abutments

Current practice is to make decks integral with the abutments. The objective is to avoid the use of joints over abutments and piers. Expansion joints are prone to leak and allow the ingress of de-icing salts into the bridge deck and substructure. In general all bridges are made continuous over intermediate supports and decks under 60 metres long with skews not exceeding 30° are made integral with their abutments.

Usually the narrow bridge is cheaper in the open abutment form and the wide bridge is cheaper in the solid abutment form. The exact transition point between the two types depends very much on the geometry and the site of the particular bridge. In most cases the open abutment solution has a better appearance and is less intrusive on the general flow of the ground contours and for these reasons is to be preferred.

It is the cost of the wing walls when related to the deck costs which swings the balance of cost in favour of the solid abutment solution for wider bridges. However the wider bridges with solid abutments produce a tunnelling effect and costs have to be considered in conjunction with the proper functioning of the structure where fast traffic is passing beneath. Solid abutments for narrow bridges should only be adopted where the open abutment solution is not possible. In the case of wide bridges the open abutment solution is to be preferred, but there are many cases where economy must be the overriding consideration.

Design Considerations

Loads transmitted by the bridge deck onto the abutment are :

i. Vertical loads from self weight of deck.

ii. Vertical loads from live loading conditions.

iii. Horizontal loads from temperature, creep movements etc. and wind.

iv. Horizontal loads from braking and skidding effects of vehicles.

These loads are carried by the bearings which are seated on the abutment bearing platform. The horizontal loads may be reduced by depending on the coefficient of friction of the bearings at the movement joint in the structure.

However, the full braking effect is to be taken, in either direction, on top of the abutment at carriageway level.

In addition to the structure loads, horizontal pressures exerted by the fill material against the abutment walls is to be considered. Also a vertical loading from the weight of the fill acts on the footing.

Vehicle loads at the rear of the abutments are considered by applying a surcharge load on the rear of the wall.

For certain short single span structures it is possible to use the bridge deck to prop the two abutments apart. This entails the abutment wall being designed as a propped cantilever.

Bearings

Bridge bearings are devices for transferring loads and movements from the deck to the substructure and foundations.

In highway bridge bearings movements are accommodated by the basic mechanisms of internal deformation (elastomeric), sliding (PTFE), or rolling. A large variety of bearings have evolved using various combinations of these mechanisms.

Design Considerations

The functions of each bearing type are :

Elastomeric: The elastomeric bearing allows the deck to translate and rotate, but also resists loads in the longitudinal, transverse and vertical directions. Loads are developed, and movement is accommodated by distorting the elastomeric pad.

Plane Sliding: Sliding bearings usually consist of a low friction polymer, polytetrafluoroethylene (PTFE), sliding against a metal plate.

This bearing does not accommodate rotational movement in the longitudinal or transverse directions and only resists loads in the vertical direction. Longitudinal or transverse loads can be accommodated by providing mechanical keys. The keys resist movement, and loads in a direction perpendicular to the keyway.

Roller: Large longitudinal movements can be accommodated by these bearings, but vertical loads only can generally be resisted.

The designer has to assess the maximum and minimum loads that the deck will exert on the bearing together with the anticipated movements (translation and rotation). Bearing manufacturers will supply a suitable bearing to meet the designers requirements.

Bearings are arranged to allow the deck to expand and contract, but retain the deck in its correct position on the substructure. A 'Fixed' Bearing does not allow translational movement. 'Sliding Guided' Bearings are provided to restrain the deck in all translational directions except in a radial direction from the fixed bearing. This allows the deck to expand and contract freely. 'Sliding' Bearings are provided for vertical support to the deck only.

Deck

Making the correct choice of deck will depend on many factors. Use the links below to find out about each type.

- *Preliminary Design - Making the correct choice of bridge deck type.*

In selecting the correct bridge type it is necessary to find a structure that will perform its required function and present an acceptable appearance at the least cost.

Decisions taken at preliminary design stage will influence the extent to which the actual structure approximates to the ideal, but so will decisions taken at detailed design stage. Consideration of each of the ideal characteristics in turn will give some indication of the importance of preliminary bridge design.

Safety: The ideal structure must not collapse in use. It must be capable of carrying the loading required of it with the appropriate factor of safety. This is more significant at detailed design stage as generally any sort of preliminary design can be made safe.

Serviceability: The ideal structure must not suffer from local deterioration/failure, from excessive deflection or vibration, and it must not interfere with sight lines on roads above or below it. Detailed design cannot correct faults induced by bad preliminary design.

Economy: The structure must make minimal demands on labour and capital; it must cost as little as possible to build and maintain. At preliminary design stage it means choosing the right types of material for the major elements of the structure, and arranging these in the right form.

Appearance: The structure must be pleasing to look at. Decisions about form and materials are made at preliminary design stage; the sizes of individual members are finalised at detailed design stage. The preliminary design usually settles the appearance of the bridge.

Constraints

The construction depth available should be evaluated. The economic implications of raising or lowering any approach embankments should then be considered.

By lowering the embankments the cost of the earthworks may be reduced, but the resulting reduction in the construction depth may cause the deck to be more expensive.

Headroom requirements have to be maintained below the deck; the minimum standards for UK Highway bridges are given in TD 27 of the Design Manual for Roads and Bridges. The Eurocode Standard (EN 1991-1-7 clause 4.3.2(1) quotes clearances from roadway surfacing to the underside of the deck to avoid impact damage.

If the bridge is to cross a road that is on a curve, then the width of the opening may have to be increased to provide an adequate site line for vehicles on the curved road.

It is important to determine the condition of the bridge site by carrying out a comprehensive site investigation. EN 1997-2: 'Ground investigation and testing' covers the requirements for the Soil Survey. Other topics which need to be considered are:

i. Existing services (Gas, Electricity, Water, etc.)

ii. Rivers and streams (liability to flood)

iii. Existing property and rights of way

iv. Access to site for construction traffic

Preliminary Design Considerations

1. A span to depth ratio of 20 will give a starting point for estimating construction depths.

2. Continuity over supports
   i. Reduces number of expansion joints.
   ii. Reduces maximum bending moments and hence construction depth or the material used.
   iii. Increases sensitivity to differential settlement.
3. Factory made units
   i. Reduces the need for soffit shuttering or scaffolding; useful when headroom is restricted or access is difficult.
   ii. Reduces site work which is weather dependent.
   iii. Dependent on delivery dates by specialist manufactures.
   iv. Specials tend to be expensive.
   v. Special permission needed to transport units of more than 29m long on the highway.
4. Length of structure
   i. The shortest structure is not always the cheapest. By increasing the length of the structure the embankment, retaining wall and abutment costs may be reduced, but the deck costs will increase.
5. Substructure
   i. The structure should be considered as a whole, including appraisal of piers, abutments and foundations. Alternative designs for piled foundations should be investigated; piling can increase the cost of a structure by up to 20%.

Costing and Final Selection

The preliminary design process will produce several apparently viable schemes. The procedure from this point is to:

i. Estimate the major quantities.
ii. Apply unit price rates - they need not be up to date but should reflect any differential variations.
iii. Obtain prices for the schemes.

The final selection will be based on cost and aesthetics. This method of costing assumes that the scheme with the minimum volume will be the cheapest, and will be true if the structure is not particularly unusual.

Reinforced Concrete

Reinforced Concrete Decks. The three most common types of reinforced concrete bridge decks are : Solid slab bridge decks are most useful for small, single or multi-span bridges and are easily adaptable for high skew. Voided slab and beam and slab bridges are used for larger, single or multi-span bridges. In circular voided decks the ratio of [depth of void] / [depth of slab] should be less than 0.79; and the maximum area of void should be less than 49% of the deck sectional area.

Analysis of Deck

For decks with skew less than 25° a simple unit strip method of analysis is generally satisfactory. For skews greater than 25° then a grillage or finite element method of analysis will be required. Skew decks develop twisting moments in the slab which become more significant with higher skew angles.

Computer analysis will produce values for Mx, My and Mxy where Mxy represents the twisting moment in the slab. Due to the influence of this twisting moment, the most economical way of reinforcing the slab would be to place the reinforcing steel in the direction of the principal moments.

However these directions vary over the slab and two directions have to be chosen in which the reinforcing bars should lie. Wood and Armer have developed equations for the moment of resistance to be provided in two predetermined directions in order to resist the applied moments Mx, My and Mxy.

Extensive tests on various steel arrangements have shown the best positions as follows

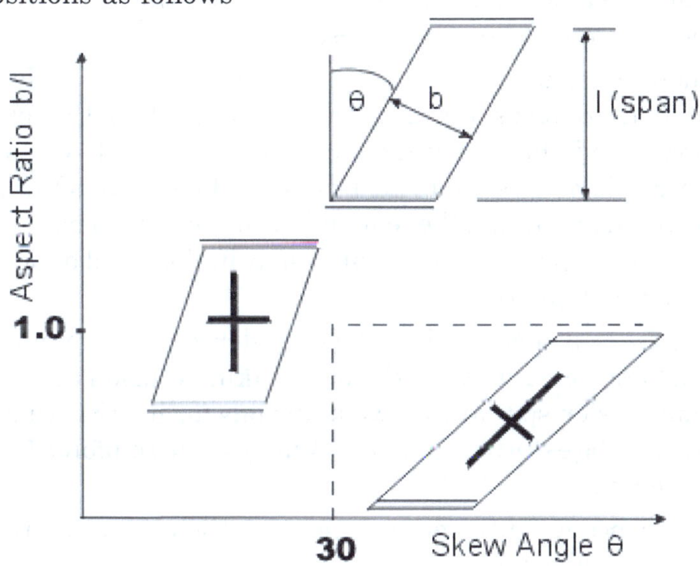

Prestressed Concrete

There are two types of deck using prestressed concrete :

i. Pre-tensioned beams with insitu concrete.

ii. Post-tensioned concrete.

The term pre-tensioning is used to describe a method of prestressing in which the tendons are tensioned before the concrete is placed, and the prestress is transferred to the concrete when a suitable cube strength is reached. Post-tensioning is a method of prestressing in which the tendon is tensioned after the concrete has reached a suitable strength. The tendons are anchored against the hardened concrete immediately after prestressing.

There are three concepts involved in the design of prestressed concrete :

i. Prestressing transforms concrete into an elastic material. By applying this concept concrete may be regarded as an elastic material, and may be treated as such for design at normal working loads. From this concept the criterion of no tensile stresses in the concrete was evolved.

   In an economically designed simply supported beam, at the critical section, the bottom fibre stress under dead load and prestress should ideally be the maximum allowable stress; and under dead load, live load and prestress the stress should be the minimum allowable stress.

   Therefore under dead load and prestress, as the dead load moment reduces towards the support, then the prestress moment will have to reduce accordingly to avoid exceeding the permissible stresses. In post-tensioned structures this may be achieved by curving the tendons, or in pre-tensioned structures some of the prestressing strands may be deflected or de-bonded near the support.

ii. Prestressed concrete is to be considered as a combination of steel and concrete with the steel taking tension and concrete compression so that the two materials form a resisting couple against the external moment. (Analogous to reinforced concrete concepts).

   This concept is utilized to determine the ultimate strength of prestressed beams.

iii. Prestressing is used to achieve load balancing. It is possible to arrange the tendons to produce an upward load which balances the downward load due to say, dead load, in which case the concrete would be in uniform compression.

Pre-tensioned Bridge Decks

Types of beams in common use are inverted T-beams, M-beams and Y beams. Inverted T-beams are generally used for spans between 7 and 16 metres and the voids between the beams are filled with insitu concrete thus forming a solid deck. M-Beams are used for spans between 14 and 30 metres and have a thin slab cast insitu spanning between the top flanges with the aim of forming a voided slab type deck. The top face of the bottom flange of M-Beams cannot be readily inspected, also the limited access makes bearing replacement difficult. As a consequence of these restrictions the Y-beam was introduced in 1990 to replace the M-beam. This lead to the production of an SY-beam which is used for spans between 32 and 40 metres.

Post-tensioned Bridge Decks

Post-tensioned bridge decks are generally composed of insitu concrete in which ducts have been cast in the required positions. When the concrete has acquired sufficient strength, the tendons are threaded through the ducts and tensioned by hydraulic jacks acting against the ends of the member. The ends of the tendons are then anchored. Tendons are then bonded to the concrete by injecting grout into the ducts after the stressing has been completed. It is possible to use pre-cast concrete units which are post-tensioned together on site to form the bridge deck.

Generally it is more economical to use post-tensioned construction for continuous structures rather than insitu reinforced concrete at spans greater than 20 metres. For simply supported spans it may be economic to use a post-tensioned deck at spans greater than 20 metres.

Composite

Composite Construction in bridge decks usually refers to the interaction between insitu reinforced concrete and structural steel.

Three main economic advantages of composite construction are:

i. For a given span and loading system a smaller depth of beam can be used than for a concrete beam solution, which leads to economies in the approach embankments.

ii. The cross-sectional area of the steel top flange can be reduced because the concrete can be considered as part of it.

iii. Transverse stiffening for the top compression flange of the steel beam can be reduced because the restraint against buckling is provided by the concrete deck.

Construction Methods

It is possible to influence the load carried by a composite deck section in a number of ways during the erection of a bridge. By propping the steel beams while the deck slab is cast and until it has gained strength, then the composite section can be considered to take the whole of the dead load. This method appears attractive but is seldom used since propping can be difficult and usually costly. With continuous spans the concrete slab will crack in the hogging regions and only the steel reinforcement will be effective in the flexural resistance, unless the concrete is prestressed.

Generally the concrete deck is 220mm to 250mm thick with beams or plate girders between 2.5m and 3.5m spacing and depths between span/20 and span/30. Composite action is developed by the transfer of horizontal shear forces between the concrete deck and steel via shear studs which are welded to the steel girder.

Steel Box Girder

Box girders have a clean, uninterrupted design line and require less maintenance because more than half of their surface area is protected from the weather. The box shape is very strong torsionally and is consequently stable during erection and in service; unlike the plate girder which generally requires additional bracing to achieve adequate stability.

The disadvantage is that box girders are more expensive to fabricate than plate girders of the same weight and they require more time and effort to design. Box girders were very popular in the late 1960's, but, following the collapse of four bridges, the Merrison Committee published design rules in 1972 which imposed complicated design rules and onerous fabrication tolerances. The design rules have now been simplified with the publication of BS5400 and more realistic imperfection limits have been set.

The load analysis and stress checks include a number of effects which are generally of second order importance in conventional plate girder design such as shear lag, distortion and warping stresses, and stiffened compression flanges. Special consideration is also required for the internal intermediate cross-frames and diaphragms at supports.

Steel Truss

Steel Truss Decks

Trusses are generally used for bridge spans between 30m and 150m where the construction depth (deck soffit to road level) is limited. The small construction depth reduces the length and height of the approach embankments that would be required for other deck forms. This can have a significant effect on the overall cost of the structure, particularly where the approach gradients cannot be steep as for railway bridges.

High fabrication and maintenance costs has made the truss type deck less popular in the UK; labour costs being relatively high compared to material costs. Where material costs are relatively high then the truss is still an economical solution. The form of construction also allows the bridge to be fabricated in small sections off site which also makes transportation easier, particularly in remote areas.

Choice of Truss

The underslung truss is the most economical as the deck provides support for the live load and also braces the compression chord. There is however the problem of the headroom clearance required under the deck which generally renders this truss only suitable for unnavigable rivers or over flood planes.

Where underslung trusses are not possible, and the span is short, it may be economical to use a half-through truss. Restraint to the compression flange is achieved by U frame action.

When the span is large, and the underslung truss cannot be used, then the through girder provides the most economic solution. Restraint to the compression flange is provided by bracing between the two top chords; this is more efficient than U frame support. The bracing therefore has to be above the headroom requirement for traffic on the deck.

Cable Stayed

Cable Stayed Decks

Cable stayed bridges are generally used for bridge spans between 150m and 1000m. They are often chosen for their aesthetics, but are generally economical for spans in excess of 250m.

Cable stayed girders were developed in Germany during the reconstruction period after the last war and attributed largely to the works of Fritz Leonhardt. Straight cables are connected directly to the deck and induce significant axial forces into the deck. The structure is consequently self anchoring and depends less on the foundation conditions than the suspension bridge.

The cables and the deck are erected at the same time which speeds up the construction time and reduces the amount of temporary works required. The cable lengths are adjusted during construction to counteract the dead load deflections of the deck due to extension in the cable.

Most early cable-stayed bridges have an orthotropic deck, mainly because the long span bridges were usually built by steel companies. It was considered economical to use composite slabs for spans up to about 250m. Developments in concrete technology have now allowed higher grade strenghs to be used. This development, combined with the increased cost of steel, has seen longer composite deck spans being used economically. Spans in excess of 600m are now being built using a steel-concrete composite box girder constuction.

Either box girders or plate girders (for the shorter spans) can be used in the deck, however if a single plane of cables is used then it is essential to use the box girder construction to achieve torsional stability.

Suspended Bridge Deck

A suspended bridge deck is a structural element of several different types of bridges. Many designs use a number of *vertical* cables or bars in tension to support the deck, these elements being attached to an overhead structure of some type:

- Through arch bridge - An overhead arch (in compression) can support a suspended deck.

Through Arch Bridge

A through arch bridge, also known as a half-through arch bridge and through-type arch bridge, is a bridge made from materials such as steel or reinforced concrete in which the base of an arch structure is below the deck, but the top rises above it, so the deck passes through the arch. Cables or beams in tension suspend the central part of the deck from the arch.

Function

For a specific construction method, especially for masonry arches, the proportions of the arch remain similar no matter the size: wider arches are thus required to be taller arches. For a semi-circular arch, the height is half of the span. Bridges across deep, narrow gorges can have their arch placed entirely beneath a flat roadway, but bridges in flatter country rise above their road approaches.

A wide bridge may require an arch so tall as to become a significant obstacle and incline for the roadway. Small bridges can be hump-backed, but larger bridges such as the Old Bridge, Pontypridd may become so steep as to require steps, making their use for wheeled traffic difficult. Railways also find arched bridges difficult as they are even less tolerant of inclines. Where simple arched bridges are used for railways on flat terrain the cost of building long approach embankments may be considerable.

Further issues are the foundations for the bridge. Arch bridges generate large side thrusts on their footings and so may require a solid bedrock foundation. Flattening the arch shape to avoid the humpback problem, such as for Brunel's Maidenhead bridge, increases this side thrust. It is often impossible to achieve a flat enough arch, simply owing to the limitations of the foundations - particularly in flat country. Historically, such bridges often became viaducts of multiple small arches.

With the availability of iron or concrete as structural materials, it became possible to construct a *through arch bridge*: a bridge where the deck does not have to be carried over the top of the arch. This requires a structure that can both support the deck from the arch by tension rods, chains or cables and that can also cantilever the deck sideways from the arch, so as to be able to pass through it. The first of these in particular cannot be achieved with masonry construction and requires wrought iron or steel.

The use of a through arch does not change the proportions or size of the arch: a large span will still require a tall arch, although this can now reach any height above the deck without obstructing traffic. The arch may also reach downwards at its sides, to either reach strong foundations or to place the roadway at a convenient height for spanning a deep valley from a plateau above. The Tyne Bridge demonstrates both of these advantages.

Notable Examples

One of the most famous bridges of this type is the Sydney Harbour Bridge in Australia, which is based on the Hell Gate Bridge in New York City, USA. Other bridges include the iconic Tyne Bridge of Newcastle upon Tyne, UK; the Bayonne Bridge that connects Staten Island, New York City to New Jersey, USA, which is longer than the Sydney Harbour Bridge, as well as the smaller, near-twin Bourne Bridge and Sagamore Bridge over the Cape Cod Canal; and the Pennybacker Bridge in Austin, Texas, USA. Wylam Railway Bridge is an early example of a through arch bridge which can be found upstream of the Tyne Bridge.

Tied-arch Bridges

Many tied-arch bridges are also through-arch bridges. As well as tying the side-loads of the arch, the tension member is also at a convenient height to form the bridge deck, as for a through-arch. The converse is not true: through-arch bridges do not imply that they are tied-arch bridges, unless they also provide the deliberate tension member that is the key to a tied-arch. Although *visually* similar, tied- and untied- through-arch bridges are quite distinct structurally and are unrelated in how they distribute their loads. In particular, cast iron bridges such as the Stanley Ferry Aqueduct may resemble tied-arch bridges, but as cast iron is weak in tension they are not structurally a tied arch.

Construction Sequence

In some locations it is not practical to support the arch from beneath during construction. In modern construction, temporary towers are erected and supported by cables anchored in the ground. Temporary cables fly from each side to support arch segments as they are constructed. When the arches are almost complete a jacking bridge is placed over or beneath the gap to force the arches apart, whence the final section is constructed in place or lifted into position.

In some cases, this type of arch has been created by constructing cantilevers from each side, with the shoreside ends bolted securely down into heavy piers. The incomplete channel ends are then constructed toward each other and either filled by construction or by lifting a prefabricated centre section. This type of construction was used in the Sydney Harbour Bridge illustrated above, with the supporting cables to the higher side of the arches removed after completion.

- Suspension bridge - A suspension cable or chain (in tension) supports a suspended deck, which can then be level (or near-level) for passage across the bridge.

Suspension

A suspension bridge is a type of bridge in which the deck (the load-bearing portion) is hung below suspension cables on vertical suspenders. Outside Tibet and Bhutan, where the first examples of this type of bridge were built in the 15th century, this type of bridge dates from the early 19th century. Bridges without vertical suspenders have a long history in many mountainous parts of the world.

This type of bridge has cables suspended between towers, plus vertical *suspender cables* that carry the weight of the deck below, upon which traffic crosses. This arrangement allows the deck to be level or to arc upward for additional clearance. Like other suspension bridge types, this type often is constructed without falsework.

The suspension cables must be anchored at each end of the bridge, since any load applied to the bridge is transformed into a tension in these main cables. The main cables continue beyond the pillars to deck-level supports, and further continue to connections with anchors in the ground. The roadway is supported by vertical suspender cables or rods, called hangers. In some circumstances, the towers may sit on a bluff or canyon edge where the road may proceed directly to the main span, otherwise the bridge will usually have two smaller spans, running between either pair of pillars and the highway, which may be supported by suspender cables or may use a truss bridge to make this connection. In the latter case there will be very little arc in the outboard main cables.

The earliest suspension bridges were ropes slung across a chasm, with a deck possibly at the same level or hung below the ropes so that the rope has a catenary shape.

Precursor

The Tibetan saint and bridge-builder Thangtong Gyalpo originated the use of iron chains in his version of early suspension bridges. In 1433, Gyalpo built eight bridges in eastern Bhutan. The only surviving chain-linked bridge of Gyalpo's was the Thangtong Gyalpo Bridge in Duksum en route to Trashi Yangtse, which was finally washed away in 2004. Gyalpo's iron chain bridges did not include a suspended deck bridge which is the standard on all modern suspension bridges today.

Instead, both the railing and the walking layer of Gyalpo's bridges used wires. The stress points that carried the screed were reinforced by the iron chains. Before the use of iron chains it is thought that Gyalpo used ropes from twisted willows or yak skins.

First

The first design for a bridge resembling the modern suspension bridge is attributed to Venetian polymath Fausto Veranzio, whose 1595 book *Machinae Novae* included drawings both for a timber and rope suspension bridge, and a hybrid suspension and cable-stayed bridge using iron chains.

The first American suspension bridge was the Jacob's Creek Bridge, built by engineer and inventor James Finley in Westmoreland County, Pennsylvania, in 1801. Finley's bridge was the first to incorporate all of the necessary components of a suspension bridge, including a suspended deck which hung by trusses. In 1808, Finley had patented his suspension bridge and by 1810, he published his design in a Philadelphia journal entitled *The Port Folio*.

Early British chain bridges included the Dryburgh Abbey Bridge (1817) and 137 m Union Bridge (1820), with spans rapidly increasing to 176 m with the Menai Suspension Bridge (1826). The Clifton Suspension Bridge (designed in 1831, completed in 1864 with a 214 m central span) is one of the longest of the parabolic arc chain type.

Wire-cable

The first wire-cable suspension bridge was the Spider Bridge at Falls of Schuylkill (1816), a modest and temporary footbridge built following the collapse of James Finley's Chain Bridge at Falls of Schuylkill (1808), shown above. The footbridge's span was 124 m, although its deck was only 0.45 m wide.

Development of wire-cable suspension bridges dates to the temporary simple suspension bridge at Annonay built by Marc Seguin and his brothers in 1822. It spanned only 18 m. The first permanent wire cable suspension bridge was Guillaume Henri Dufour's Saint Antoine Bridge in Geneva of 1823, with two 40 m spans. The first with cables assembled in mid-air in the modern method was Joseph Chaley's Grand Pont Suspendu in Fribourg, in 1834.

In the United States, the first major wire-cable suspension bridge was the Wire Bridge at Fairmount in Philadelphia, Pennsylvania. Designed by Charles Ellet, Jr. and completed in 1842, it had a span of 109 m. Ellet's Niagara Falls Suspension Bridge (1847–48) was abandoned before completion. It was used as scaffolding for John A. Roebling's double decker railroad and carriage bridge (1855).

Structural Behaviour

Structural Analysis

The main forces in a suspension bridge of any type are tension in the cables and compression in the pillars. Since almost all the force on the pillars is vertically downwards and they are also stabilized by the main cables, the pillars can be made quite slender, as on the Severn Bridge, on the Wales-England border.

In a suspended deck bridge, cables suspended via towers hold up the road deck. The weight is transferred by the cables to the towers, which in turn transfer the weight to the ground.

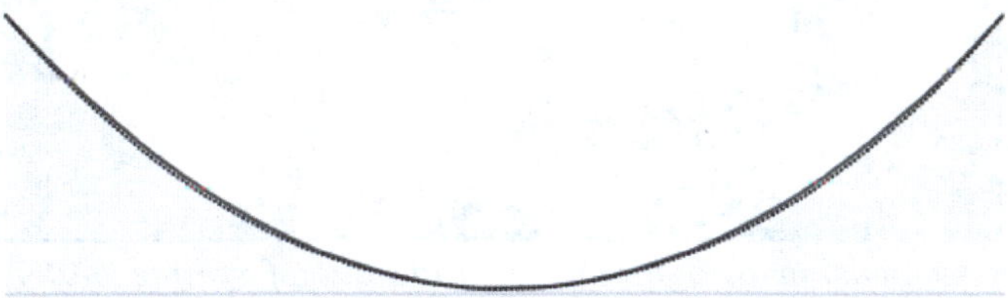

Figure: *Comparison of a catenary (black dotted curve) and a parabola (red solid curve) with the same span and sag. The catenary represents the profile of a simple suspension bridge, or the cable of a suspended-deck suspension bridge on which its deck and hangers have negligible mass compared to its cable. The parabola represents the profile of the cable of a suspended-deck suspension bridge on which its cable and hangers have negligible mass compared to its deck. The profile of the cable of a real suspension bridge with the same span and sag lies between the two curves.*

Assuming a negligible weight as compared to the weight of the deck and vehicles being supported, the main cables of a suspension bridge will form a parabola (very similar to a catenary, the form the unloaded cables take before the deck is added).

One can see the shape from the constant increase of the gradient of the cable with linear (deck) distance, this increase in gradient at each connection with the deck providing a net upward support force. Combined with the relatively simple constraints placed upon the actual deck, this makes the suspension bridge much simpler to design and analyze than a cable-stayed bridge, where the deck is in compression.

Advantages

Figure: A suspension bridge can be made out of simple materials such as wood and common wire rope.

- Longer main spans are achievable than with any other type of bridge
- Less material may be required than other bridge types, even at spans they can achieve, leading to a reduced construction cost
- Except for installation of the initial temporary cables, little or no access from below is required during construction, for example allowing a waterway to remain open while the bridge is built above
- May be better to withstand earthquake movements than heavier and more rigid bridges

Disadvantages

- Considerable stiffness or aerodynamic profiling may be required to prevent the bridge deck vibrating under high winds
- The relatively low deck stiffness compared to other (non-suspension) types of bridges makes it more difficult to carry heavy rail traffic where high concentrated live loads occur

- Some access below may be required during construction, to lift the initial cables or to lift deck units. This access can often be avoided in cable-stayed bridge construction

Variations

Underspanned

Figure: *Eyebar chain cables of Clifton Suspension Bridge*

Figure: *The Yichang Bridge, a plate deck suspension bridge, over the Yangtze River in China*

In an underspanned suspension bridge, the main cables hang entirely below the bridge deck, but are still anchored into the ground in a similar way to the conventional type. Very few bridges of this nature have been built, as the deck is inherently less stable than when suspended below the cables. Examples include the Pont des Bergues of 1834 designed by Guillaume Henri Dufour; James Smith's Micklewood Bridge; and a proposal by Robert Stevenson for a bridge over the River Almond near Edinburgh.

Roebling's Delaware Aqueduct (begun 1847) consists of three sections supported by cables. The timber structure essentially hides the cables; and from a quick view, it is not immediately apparent that it is even a suspension bridge.

Suspension Cable Types

The main suspension cable in older bridges was often made from chain or linked bars, but modern bridge cables are made from multiple strands of wire. This contributes greater redundancy; a few flawed strands in the hundreds used pose very little threat, whereas a single bad link or eyebar can cause failure of the entire bridge. (The failure of a single eyebar was found to be the cause of the collapse of the Silver Bridge over the Ohio River). Another reason is that as spans increased, engineers were unable to lift larger chains into position, whereas wire strand cables can be largely prepared in mid-air from a temporary walkway.

Deck Structure Types

Most suspension bridges have open truss structures to support the roadbed, particularly owing to the unfavourable effects of using plate girders, discovered from the Tacoma Narrows Bridge (1940) bridge collapse. Recent developments in bridge aerodynamics have allowed the re-introduction of plate structures. In the picture of the Yichang Bridge, note the very sharp entry edge and sloping undergirders in the suspension bridge shown. This enables this type of construction to be used without the danger of vortex shedding and consequent aeroelastic effects, such as those that destroyed the original Tacoma Narrows bridge.

Forces

Three kinds of forces operate on any bridge: the dead load, the live load, and the dynamic load. Dead load refers to the weight of the bridge itself. Like any other structure, a bridge has a tendency to collapse simply because of the gravitational forces acting on the materials of which the bridge is made. Live load refers to traffic that moves across the bridge as well as normal environmental factors such as changes in temperature, precipitation, and winds. Dynamic load refers to environmental factors that go beyond normal weather conditions, factors such as sudden gusts of wind and earthquakes. All three factors must be taken into consideration when building a bridge.

Use Other Than Road and Rail

Figure: *Cable-suspended footbridge at Dallas Fort Worth Airport Terminal D*

The principles of suspension used on the large scale may also appear in contexts less dramatic than road or rail bridges. Light cable suspension may prove less expensive and seem more elegant for a footbridge than strong girder supports. Where such a bridge spans a gap between two buildings, there is no need to construct special towers, as the buildings can anchor the cables. Cable suspension may also be augmented by the inherent stiffness of a structure that has much in common with a tubular bridge.

Construction Sequence (Wire Strand Cable Type)

Typical suspension bridges are constructed using a sequence generally described as follows.

Figure: Suspender cables and suspender cable band on the Golden Gate Bridge in San Francisco. Main cable diameter is 36 inches, and suspender cable diameter is $3^1/_2$ inches.

Depending on length and size, construction may take anywhere between a year and a half (construction on the original Tacoma Narrows Bridge took only 19 months) to as many as a decade (the Akashi-Kaikyô Bridge's construction began in May 1986 and was opened in May, 1998 – a total of twelve years).

1. Where the towers are founded on underwater piers, *caissons* are sunk and any soft bottom is excavated for a foundation. If the bedrock is too deep to be exposed by excavation or the sinking of a caisson, pilings are driven to the bedrock or into overlying hard soil, or a large concrete pad to distribute the weight over less resistant soil may be constructed, first preparing the surface with a bed of compacted gravel. (Such a pad footing can also accommodate the movements of an

active fault, and this has been implemented on the foundations of the cable-stayed Rio-Antirio bridge. The piers are then extended above water level, where they are capped with pedestal bases for the towers.

***Figure:** Lions' Gate Bridge with deck under construction from the span's centre*

2. Where the towers are founded on dry land, deep foundation excavation or pilings are used.
3. From the tower foundation, towers of single or multiple columns are erected using high-strength reinforced concrete, stonework, or steel. Concrete is used most frequently in modern suspension bridge construction due to the high cost of steel
4. Large devices called *saddles*, which will carry the main suspension cables, are positioned atop the towers. Typically of cast steel, they can also be manufactured using riveted forms, and are equipped with rollers to allow the main cables to shift under construction and normal loads.
5. *Anchorages* are constructed, usually in tandem with the towers, to resist the tension of the cables and form as the main anchor system for the entire structure. These are usually anchored in good quality rock, but may consist of massive reinforced concrete deadweights within an excavation. The anchorage

structure will have multiple protruding open *eyebolts* enclosed within a secure space.

6. Temporary suspended walkways, called *catwalks*, are then erected using a set of guide wires hoisted into place via winches positioned atop the towers. These catwalks follow the curve set by bridge designers for the main cables, in a path mathematically described as a catenary arc. Typical catwalks are usually between eight and ten feet wide, and are constructed using wire grate and wood slats.
7. Gantries are placed upon the catwalks, which will support the main cable spinning reels. Then, cables attached to winches are installed, and in turn, the main cable spinning devices are installed.
8. High strength wire (typically 4 or 6 gauge galvanized steel wire), is pulled in a loop by pulleys on the traveler, with one end affixed at an anchorage. When the traveller reaches the opposite anchorage the loop is placed over an open anchor eyebar. Along the catwalk, workers also pull the cable wires to their desired tension. This continues until a bundle, called a "cable strand" is completed, and temporarily bundled using stainless steel wire. This process is repeated until the final cable strand is completed. Workers then remove the individual wraps on the cable strands (during the spinning process, the shape of the main cable closely resembles a hexagon), and then the entire cable is then compressed by a travelling hydraulic press into a closely packed cylinder and tightly wrapped with additional wire to form the final circular cross section. The wire used in suspension bridge construction is a galvanized steel wire that has been coated with corrosion inhibitors.
9. At specific points along the main cable (each being the exact distance horizontally in relation to the next) devices called "cable bands" are installed to carry steel wire ropes called *Suspender cables.* Each suspender cable is engineered and cut to precise lengths, and are looped over the cable bands. In some bridges, where the towers are close to or on the shore, the suspender cables may be applied only to the central span. Early suspender cables were fitted with zinc jewels and a set of steel washers, which formed the support for the deck. Modern suspender cables carry a shackle-type fitting.

10. Special lifting hoists attached to the suspenders or from the main cables are used to lift prefabricated sections of bridge deck to the proper level, provided that the local conditions allow the sections to be carried below the bridge by barge or other means. Otherwise, a travelling cantilever derrick may be used to extend the deck one section at a time starting from the towers and working outward.

    If the addition of the deck structure extends from the towers the finished portions of the deck will pitch upward rather sharply, as there is no downward force in the centre of the span.

    Upon completion of the deck the added load will pull the main cables into an arc mathematically described as a parabola, while the arc of the deck will be as the designer intended – usually a gentle upward arc for added clearance if over a shipping channel, or flat in other cases such as a span over a canyon. Arched suspension spans also give the structure more rigidity and strength.
11. With completion of the primary structure various details such as lighting, handrails, finish painting and paving are installed or completed.

Longest Spans

Suspension bridges are typically ranked by the length of their main span. These are the ten bridges with the longest spans, followed by the length of the span and the year the bridge opened for traffic:

1. Akashi Kaikyô Bridge (Japan), 1991 m – 1998
2. Xihoumen Bridge (China), 1650 m – 2009
3. Great Belt Bridge (Denmark), 1624 m – 1998
4. Yi Sun-sin bridge (South Korea), 1545 m – 2012
5. Runyang Bridge (China), 1490 m – 2005
6. Fourth Nanjing Yangtze Bridge (China), 1418 m – 2012
7. Humber Bridge (England, United Kingdom), 1410 m – 1981 (longest span from 1981 until 1998)
8. Jiangyin Suspension Bridge (China), 1385 m – 1997
9. Hardanger Bridge (Norway), 1380 m - 2013
10. Tsing Ma Bridge (Hong Kong, China), 1377 m – 1997 (longest span with both road and metro)

Drainage

Bridge drainage covers the collection and removal of waters from a bridge deck. To accomplish this function, drains are placed adjacent to curbs for collection of water which is then either dumped directly on the ground or is conveyed to a suitable disposal point.

Bridge Deck Drainage is Divided into Two Classes

- Class I. All locations, usually urban, where drainage must be carried via piping to some suitable disposal point. Short vertical or near vertical down spouts to clear girder flanges are not considered in this category.
- Class II. All locations, usually rural, where drainage may be disposed of by free fall directly under the drain.

Four Standard Types of Structure Drains

- Type A. To be used for Class II drainage at locations adjacent to median curbs where a grate is desirable.
- Type B. To be used for Class II drainage at locations adjacent to barrier rail curbs.
- Type C. To be used for both Class I and Class II drainage. Intended primarily for use on steel or precast girder construction in deck areas where there is no main reinforcing running longitudinally in the deck to conflict with a drain of this size. This drain will intercept more deck runoff than other types.
- Type D. To be used for Class I Drainage and is the type most commonly used for this class.

D-1 has a deep basin and is preferred when drains need to be located in interior bays of box girder bridges.

D-2 has a shallow basin and is intended for use in deck overhangs. Make sure a D-3 cannot be used before calling out this drain.

D-3 is an improvement over the D-2 drain and is the preferred drain for deck overhangs

Structure Drains for Special Applications

Double Wide Type D Drain: When hydraulic demand exceeds the capacity of a single Type D drain inlet, consider using a double wide Type D drain.

Slotted Drains: Slotted drains in bridge decks should be used only to intercept sheet- flow. Drains should be located in the shoulder either normal to traffic or parallel to traffic, 12" from edge of outside traffic lane. For wide bridges with flat cross slopes, additional drains may be located 12" from traffic lane lines.

Drop-Thru Drains and Scuppers: A good, economical choice for Class II drainage. Locate these drains so water and roadway debris will not fall on pedestrians, motorists, and railroad right-of-ways. Inform the District when considering this type of drain as there may be additional restrictions that would prohibit scuppers and drop-thru drains.

Drain Inlet Location

It is the policy of Structure Design to provide drains at the following locations:

a. At low points where there is a sag vertical curve.

b. Where ponding may occur due to a combination of "near-flat" profile grades and initial residual camber.

c. Adjacent to any area where water will cross the deck due to reversal of superelevation.

d. The termination of a curb nosing which separates two roadways.

e. Immediately preceding and upstream of expansion joint seal assemblies in bridge decks. Complete interception of all flow prior to the joint is not required.

Drainage Disposal

Drainage disposal, required for Class I drainage only, is accomplished in one of two ways via concealed piping by either:

a. Discharging at ground level adjacent to a gutter or other disposal area; or

b. Discharging underground into a catch basin or connecting to a storm drain.

Coordinate with the District to Determine the Method of Discharge.

Criteria for Piping

a. Exposed piping detracts from the appearance of the structure and should be avoided.

b. Piping should be 6" minimum diameter welded steel pipe with 1/8" minimum wall thickness.

For multiple inlets, near-horizontal slopes, or runs of 50' or more, 8" piping is preferred.

Runs should be on as steep a grade as conditions will allow with a maximum of 4 inlets per run. Inlets should be located upgrade from the bent or abutment outlets whenever possible.

All bends should be smooth and on an 18" minimum radius for 6" diameter pipe and a 24" minimum radius for 8" diameter pipe. Mitered bends shall not be used. The length of near horizontal (2 percent minimum) runs should be kept to a minimum.

c. Expansion couplings must be shown on the plans where drain pipes cross expansion joints at hinges and abutments.

a. The use of clean out openings must be given careful consideration by the designer.

When the piping system consists of a single run from inlet to outlet, cleaning is usually done from the outlet end with a power driven plumber's auger. For this situation, a cleanout opening should not be used, because it may provide a "blind alley" for the auger and make cleaning from the outlet impossible. For a more complex piping system, properly placed cleanout openings can facilitate cleaning by providing additional access points in the line.

Footing/Foundations

A foundation is the lowest and supporting layer of a structure. Foundations are generally divided into two categories: shallow foundations and deep foundations.

Earthfast or Post in Ground Construction

Buildings and structures have a long history of being built with wood in contact with the ground. Post in ground construction may technically have no foundation. Timber pilings were used on soft or wet ground even below stone or masonry walls. In marine construction and bridge building a crisscross of timbers or steel beams in concrete is called grillage.

Padstones

Perhaps the simplest foundation is the padstone, a single stone which both spreads the weight on the ground and raises the timber off of the ground. Staddle stones are a specific type of padstone.

Stone Foundations

Dry laid stone foundations are common in many parts of the world.

Modern Foundation Types

Shallow Foundations

Shallow foundations, often called footings, are usually embedded about a metre or so into soil. One common type is the spread footing which consists of strips or pads of concrete (or other materials) which extend below the frost line and transfer the weight from walls and columns to the soil or bedrock.

Another common type of shallow foundation is the slab-on-grade foundation where the weight of the building is transferred to the soil through a concrete slab placed at the surface. Slab-on-grade foundations can be reinforced mat slabs, which range from 25 cm to several metres thick, depending on the size of the building, or post-tensioned slabs, which are typically at least 20 cm for houses, and thicker for heavier structures.

Deep Foundations

A deep foundation is used to transfer the load of a structure down through the upper weak layer of topsoil to the stronger layer of subsoil below. There are different types of deep footings including impact driven piles, drilled shafts, caissons, helical piles, geo-piers and earth stabilized columns. The naming conventions for different types of footings vary between different engineers. Historically, piles were wood, later steel, reinforced concrete, and pre-tensioned concrete.

Monopile Foundation

A monopile foundation is a type of deep foundation which uses a single, generally large-diameter, structural element embedded into the earth to support all the loads (weight, wind, etc.) of a large above-surface structure.

A large number of monopile foundations have been utilized in recent years for economically constructing fixed-bottom offshore wind farms in shallow-water subsea locations. For example, a single wind farm off the coast of England went online in 2008 with over 100 turbines, each mounted on a 4.7-metre-diameter monopile footing in ocean depths up to 18 metres of water.

Design

Foundations are designed to have an adequate load capacity with limited settlement by a geotechnical engineer, and the footing itself may be designed structurally by a structural engineer.

The primary design concerns are settlement and bearing capacity. When considering settlement, total settlement and differential settlement is normally considered. Differential settlement is when one part of a foundation settles more than another part. This can cause problems to the structure the foundation is supporting.

Joints

A joint is a junction where building elements meet without applying a static load from one element to another. When one or more of these vertical or horizontal elements that meet are required by the local building code to have a fire-resistance rating, the resulting opening that makes up the joint must be firestopped in order to restore the required compartmentalisation.

Such joints are often subject to movement, as a function of the building's design basis. Firestops must be able to demonstrate the ability to withstand operational movement prior to fire testing.

Whether or not the building elements forming the joint have a fire-resistance rating, the joint design must still consider the anticipated operational movement of each joint. Timing is also important, as freshly poured concrete shrinks particularly during the first few months of a new building, potentially causing joint size changes.

Perimeter Slab Edge

Firestopping at the "perimeter slab edge", which is gaps between the floors and the backpan of the curtain wall or precast concrete panels, slow the passage of fire and combustion gases between floors. The firestop at the perimeter slab edge is considered a continuation of the fire-resistance rating of the floor slab. The curtain wall itself, however, is not ordinarily required to have a rating. Fire compartmentalisation is based upon compartments enclosed on all sides to avoid fire and smoke migration beyond each engaged compartment.

A non-fire-resistance rated curtain wall prevents the complete compartment (or envelope) from being fire-resistance rated. The provision of firestopping at the joint of the floor slab and curtain wall can still add to the overall fire safety of the building by slowing the advance of the fire in the building interior from floor to floor. The use of fire sprinklers has been shown to mitigate the spread of fire by keeping the fire from growing to a size large enough to cause structural damage and break the exterior glass. If the building is not equipped with sprinklers, the fire can travel up the outside of the curtain wall if the glass on the floor of fire origin is shattered due to radiant heat, causing flames to lick up the outside of the building, resulting in the glass in floors above to break. Falling glass can endanger pedestrians, firefighters and firehoses below. An example of mode of fire spread is the First Interstate Bank Fire in Los Angeles, California. This building is now the Aon Centre.

The fire leapfrogged up the tower by shattering the glass. The exterior curtain wall aluminium skeleton holding the glass failed. Exceptionally sound cementitious spray fireproofing also helped to delay and ultimately to avoid the possible collapse of the building, by preventing the steel building structure from reaching the critical temperature at which steel weakens and can no longer hold its intended load. This fire proved the positive, collective effect of both active fire protection (sprinklers) to control fire size in its incipient stages, and passive fire protection (fireproofing) if the fire sprinklers are not effective.

Head-of-Wall (HOW)

Where fire-resistance rated wall assemblies, be they masonry or drywall, meet the underside of the floor slab above, a movement joint results, which can be subject to compression, as the freshly placed concrete cures and shrinks all over a new building. This joint must be firestopped in a flexible manner. For concrete block walls that are not load-bearing, the contractor is often instructed to stop the wall about 25mm short of the slab above. The drywaller has to fasten a track to the underside of the slab above, which holds the metal studs. The drywall is then either stopped short allowing for a flexible buffer to be inserted or caulked in place, or one may use purpose-made tracks that hold the studs but permit up and down movement without damage to the system.

Parapet

A parapet is a barrier which is an extension of the wall at the edge of a roof, terrace, balcony, or other structure. Where extending above a roof, it may simply be the portion of an exterior wall that continues above the line of the roof surface, or may be a continuation of a vertical feature beneath the roof such as a fire wall or party wall. Parapets were originally used to defend buildings from military attack, but today they are primarily used as guard rails and to prevent the spread of fires.

Building Parapets

Figure: *Elaborate parapets flank the gable of King's College Chapel, Cambridge.*

Figure: *A modern parapet with electric lights functions as a guard rail along the BP Pedestrian Bridge in Chicago.*

The word comes ultimately from the Italian *parapetto* (*parare* = to cover/defend and *petto* = breast). The German term *Brustwehr* has the same significance.

Parapets may be plain, embattled, perforated or panelled, which are not mutually exclusive terms.

- Plain parapets are upward extensions of the wall, sometimes with a coping at the top and corbel below.
- Embattled parapets may be panelled, but are pierced, if not purely as stylistic device, for the discharge of defensive projectiles.
- Perforated parapets are pierced in various designs such as circles, trefoils, or quatrefoils.
- Panelled parapets are ornamented by a series of panels, either oblong or square, and more or less enriched, but not perforated. These are common in the Decorated and Perpendicular periods.

Parapet Roofs

Parapets surrounding roofs are extremely common in London. This dates from the Building Act of 1707 which banned projecting wooden eaves in the cities of Westminster and London as a fire risk. Instead an 18-inch brick parapet was required, with the roof set behind. This was continued in many Georgian houses, as it gave the appearance of a flat roof which accorded with the desire for classical proportions.

The Mosaic law also prescribed parapets for newly constructed houses as a safety measure (Deuteronomy 22:8).

Fire Wall Parapets

Many firewalls are required to have a parapet, a portion of the wall extending above the roof. The parapet is required to be as fire resistant as the lower wall, and extend a distance prescribed by building code.

Bridge Parapets

Parapets on bridges and other highway structures (such as retaining walls) prevent users from falling off where there is a drop. They may also be meant to restrict views, to prevent rubbish passing below, and to act as noise barriers.

Bridge parapets may be made from any material, but structural steel, aluminium, timber and reinforced concrete are common. They may be of solid or framed construction. In European standards, parapets are defined as a sub-category of "vehicle restraint systems" or "pedestrian restraint systems".

Parapets in Fortification

In terms of fortification, a parapet (or breastwork) is a wall of stone, wood or earth on the outer edge of a defensive wall or trench, which shelters the defenders. In medieval castles, they were often crenellated. In later artillery forts, parapets tend to be higher and thicker. They could be provided with embrasures for the fort's guns to fire through, and a banquette or fire-step so that defending infantry could shoot over the top. The top of the parapet often slopes towards the enemy to enable the defenders to shoot downwards; this incline is called the *superior talus*.

Piers

A pier is a raised structure, including bridge and building supports and walkways, typically supported by widely spread piles or pillars. The lighter structure of a pier allows tides and currents to flow almost unhindered, whereas the more solid foundations of a quay or the closely spaced piles of a wharf can act as a breakwater, and are consequently more liable to silting. Piers can range in size and complexity from a simple lightweight wooden structure to major structures extended over 1600 metres. In American English, pier may be synonymous with dock.

Piers have been built for several purposes, and because these different purposes have distinct regional variances, the term *pier* tends to have different nuances of meaning in different parts of the world. Thus in North America and Australia, where many ports were, until recently, built on the multiple pier model, the term tends to imply a current or former cargo-handling facility. In Europe in contrast, where ports more often use basins and river-side quays than piers, the term is principally associated with the image of a Victorian cast iron pleasure pier. However, the earliest piers pre-date the Victorian age.

Types of Pier

Piers can be categorized into different groupings according to the principal purpose. However there is considerable overlap between these categories. For example, pleasure piers often also allow for the docking of pleasure steamers and other similar craft, whilst working piers have often been converted to leisure use after being rendered obsolete by advanced developments in cargo-handling technology.

Working Piers

Working piers were built for the handling of passengers and cargo onto and off ships or (as at Wigan Pier) canal boats. Working piers themselves fall into two different groups. Longer individual piers are often found at ports with large tidal ranges, with the pier stretching far enough off shore to reach deep water at low tide. Such piers provided an economical alternative to impounded docks where cargo volumes were low, or where specialist bulk cargo was handled, such as at coal piers.

The other form of working pier, often called the finger pier, was built at ports with smaller tidal ranges. Here the principal advantage was to give a greater available quay length for ships to berth against compared to a linear littoral quayside, and such piers are usually much shorter. Typically each pier would carry a single transit shed the length of the pier, with ships berthing bow or stern in to the shore. Some major ports consisted of large numbers of such piers lining the foreshore, classic examples being the Hudson River frontage of New York, or the Embarcadero in San Francisco.

Figure: *This pier deals with varying water levels, ships and freeboards by means of stairs.*

The advent of container shipping, with its need for large container handling spaces adjacent to the shipping berths, has made working piers obsolete for the handling of general cargo, although some still survive for the handling of passenger ships or bulk cargos. One example, is in use in Progreso, Yucatán, where a pier extends more than 4 miles into the Gulf of Mexico, making it the longest pier in the world. The Progreso Pier supplies much of the peninsula with transportation for the fishing and cargo industries and serves as a port for large cruise ships in the area. Many other working piers have been demolished, or remain derelict, but some have been recycled as pleasure piers. The best known example of this is Pier 39 in San Francisco.

At Southport and the Tweed River on the Gold Coast in Australia, there are piers that support equipment for a sand bypassing system that maintains the health of sandy beaches and navigation channels.

Pleasure Piers

***Figure:** Print of Victorian pier in Margate, 1897*

Pleasure piers were first built in England, during the 19th century. The earliest structures were Ryde Pier, built in 1813/4, Leith Trinity Chain Pier, built in 1821, and Brighton Chain Pier, built in 1823. Only the oldest of these piers still remains. At that time the introduction of the railways for the first time permitted mass tourism to dedicated seaside resorts. However, the large tidal ranges at many such resorts meant that for much of the day, the sea was not visible from dry land. The pleasure pier was the resorts' answer, permitting holiday makers to promenade over and alongside the sea at all times. The longest Pleasure pier in the world is at Southend-on-sea, Essex, and extends 2,158 metres (1.34 mi) into the Thames estuary. With a length of 2,745 feet (836.68 m), the longest pier on the West Coast of the United States is the Santa Cruz Wharf.

Pleasure piers often include other amusements and theatres as part of the attraction. Such a pier may be open air, closed, or partly open, partly closed. Sometimes a pier has two decks. Early pleasure piers were of wooden construction, with iron structures being introduced with the construction in 1855 of Margate Jetty, in Margate, England. Margate was wrecked in storms in 1978 and was never repaired. The oldest iron pleasure pier still remaining is in Southport, England, and dates from 1860 - however the world's oldest iron pier dates from 1834 and is in Gravesend, Kent. Gravesham council have recently purchased and refurbished this passenger pier.

Fishing Piers

Many piers are built for the purpose of providing boatless anglers access to fishing grounds that are otherwise inaccessible.

Piers of the World

Figure: *At 6.5km (4 miles) the remote terminal in Progreso, Yucatán, Mexico is the largest pier in the world.*

England and Wales

The first recorded pier in England was Ryde Pier, opened in 1814 on the Isle of Wight, as a landing stage to allow ferries to and from the mainland to berth. It is still used for this purpose today. However it has had a leisure function in the past, with the pier head once containing a pavilion. There are still refreshment facilities today. The oldest cast iron pier in the world is Gravesend Town Pier, in Kent, which opened in 1834. However, it is not recognised by the National Piers Society as being a seaside pier.

In their heyday, there were many pleasure piers across England and Wales. These were found in most fashionable seaside resorts during the Victorian era. There are still a significant number of piers of architectural merit still standing, although some have been lost. The most well known piers are perhaps the two at Brighton in East Sussex and the three at Blackpool in Lancashire. Two piers, Brighton's now derelict West Pier and Clevedon Pier, were Grade 1 listed:Brighton West lost its status after a series of fires and storms. The Birnbeck Pier in Weston-super-Mare is the only pier in the world that is linked to an island. The National Piers Society gives a figure of 55 surviving seaside piers in England and Wales.

Netherlands

Scheveningen, the coastal resort town of The Hague, boasts the largest pier in the Netherlands, completed in 1961. A crane, built on top of the pier's panorama tower, provides the opportunity to make a 60-metre (200 ft) high bungee jump over the North Sea waves. The present pier is a successor of an earlier pier, which was completed in 1901 but in 1943 destroyed by the German occupation forces.

Belgium

In Blankenberge a first pleasure pier was built in 1894. After its destruction in the World War I, a new pier was built in 1933. It remained till the present day, but was partially transformed and modernized in 1999–2004.

In Nieuwpoort, Belgium there is a pleasure pier on both sides of the river IJzer.

Wing Walls

A wing wall is a smaller wall attached or next to a larger wall or structure. In a bridge, the wing walls are adjacent to the abutments and act as retaining walls. They are generally constructed of the same material as those of abutments. The wing walls can either be attached to the abutment or be independent of it. Wing walls are provided at both ends of the abutments to retain the earth filling of the approaches. Their design period depends upon the nature of the embankment and does not depend upon the type or parts of the bridge.

The soil and fill supporting the roadway and approach embankment are retained by the wing walls, which can be at a right angle to the abument or splayed at different angles. The wing walls are generally constructed at the same time and of the same materials as the abuments.

Classification of Wing Walls

Wing walls can be classified according to their position in plan with respect to banks and abutments. The classification is as follows:

1. Straight Wing walls: They are used for small bridges, on drains with low banks and for railway bridges in cities (weep holes are provided).
2. Splayed Wing walls: These are used for bridges across rivers. They provide smooth entry and exit to the water. The splay is usually 45°. Their top width is 0.5 m, face batter 1 in 12 and back batter 1 in 6, weep holes are provided.
3. Return Wing walls: They are used where banks are high and hard or firm. Their top width is 1.5 m and face is vertical and back battered 1 in 4. Scour can be a problem for wing walls and abutments both, as the water in the stream erodes the supporting soil.

Other Uses

Wing walls provide smooth entry of water into the bridge site and provide support and protect the embankment. Wing walls can serve as buttresses to support walls. They can also be purely decorative.

Seismic Retrofit Technology

Retrofitting of existing structures with insufficient seismic resistance accounts for a major portion of the total cost of hazard mitigation. Thus, it is of critical importance that the structures that need seismic retrofitting are identified correctly, and an optimal retrofitting is conducted in a cost effective fashion. Once the decision is made, seismic retrofitting can be performed through several methods with various objectives such as increasing the load, deformation, and/ or energy dissipation capacity of the structure. Conventional as well as emerging retrofit methods are briefly presented in the following subsections.

Conventional Strengthening Methods

Conventional retrofitting methods include addition of new structural elements to the system and enlarging the existing members. Addition of shear walls and bracings.

Retrofit of Structures Using Innovative Materials

Current research on advanced materials in civil engineering is mainly concentrated on high performance concrete and steel, and fibre reinforced plastic (FRP) composites.

FRP composite materials have experienced a continuous increase of use in structural strengthening and repair applications around the world in the last fifteen years. High specific stiffness and specific weight combined with superior environmental durability of these materials have made them a competing alternative to the conventional strengthening methods. It was shown through experimental and analytical studies that externally bonded FRP composites can be applied to various structural members including columns, beams, slabs, and walls to improve their structural performance such as stiffness, load carrying capacity, and ductility.

FRP composites have enjoyed varying degrees of success in different types of applications. In general, applications that allow complete wrapping of the member with FRP have proven to be effective. Wrapping of columns to increase their load and deformation capacity is the most effective and most commonly used method of retrofitting with composites. However, certain performance and failure mode issues regarding different wrapping configuration and fibre orientations, still need to be well understood. When wrapping is difficult or not allowed, such as when strengthening beams, slabs, or walls, success of the method is sometimes hindered by premature debonding failures.

Such problems can be reduced through proper design and anchorage of the external FRP reinforcement. Thus, decision makers must approach using these materials with caution and must ensure that the design is performed with adequate knowledge and skill, and verified through laboratory testing.

Limited research and applications regarding seismic retrofitting of building systems with FRP composites have shown that composites retrofitting does not significantly alter the stiffness and dynamic properties of the building. The main benefit of retrofitting with composites is the increase in deformation capacity of the building, and in its load capacity to an extent. This may achievethe retrofit objectives for buildings with lightly insufficient seismic resistance. For buildings with large seismic deficiencies, a combination of conventional and FRP strengthening techniques may prove to be an effective retrofitting solution.

As a practical and economical solution, the retrofit design involved replacement of the existing window frames with structural steel frames constructed from steel C-sections. A verification analysis of the retrofitted building showed that installation of steel window frames largely decreased the stress concentrations, but did not suffice to reduce all stresses to acceptable levels. For this reason, additional retrofitting was designed using externally bonded FRP composites around the openings in the walls to prevent or delay concrete crack propagation by bridging the stresses at crack locations. Thus, by combining conventional and innovative materials, an effective and economical retrofit design was achieved that did not significantly interfere with the function or historical and architectural character of the building.

FRP composites are widely recognized for their potential use in seismic retrofitting applications. To achieve wide-range use of these materials, however, there is need for further research into a number of issues related to mechanics, design, and durability of FRP retrofitted concrete and steel systems.

Despite considerable progress in these areas since early last decade (ACI-440F, 2000), further improvements are necessary to meet the needs of the retrofit industry. Failure mechanisms, with emphasis on brittle shear and debonding failures, must be thoroughly understood and associated design procedures must be incorporated in design codes. Influence of cyclic and fatigue loading on the FRP strengthened member performance must be characterized and accounted for in the design process.

Although FRP composites are known for their favourable durability characteristics, only limited information is available on long-term durability and performance of FRP bonded concrete and steel systems.

These issues need to be investigated through accelerated test studies and related design, application and protection requirements must be included in the design codes.

Base Isolation

The seismic base isolation technology involves placing flexible isolation systems between the foundation and the superstructure. By means of their flexibility and energy absorption capability, the isolation systems reflect and absorb part of the earthquake input energy before this energy is fully transmitted to the superstructure, reducing the energy dissipation demand on the superstructure. Base isolation causes the natural period of the structure to increase and results in increased displacements across the isolation level and reduced accelerations and displacements in the superstructure during an earthquake. This not only provides safety against collapse, but also largely reduces damage, which is crucial for facilities that should remain operational after severe earthquakes such as emergency response centres, hospitals, and fire stations. Base isolation can also be used in seismic retrofitting of historic structures without impairing their architectural characteristics by reducing the induced seismic forces.

Base isolation is generally suitable for low to medium rise buildings, usually up to 10-12 stories high, which have their fundamental frequencies in the range of expected dominant frequencies of earthquakes. Superstructure characteristics such as height, width, aspect ratio, and stiffness are important in determining the applicability and effectiveness of seismic isolation. The seismicity of the region and the underlying soil conditions should also be considered in the feasibility studies and design process. Base isolation should be avoided in areas where expected fundamental frequencies of the earthquakes are in the lower frequency domain or on soft soil sites where amplification of low earthquake frequencies may occur. One other constraint in the application of base isolation is the large relative displacements between the superstructure and the supporting ground at the isolation level. A clearance around the building must be provided and maintained through the life of the structure to accommodate the expected large displacements. Such displacements may be reduced with the incorporation of additional stiffness and energy dissipation mechanisms in the isolation system.

The International Building Code (IBC, 2000) and FEMA 356 (FEMA, 2000) specify the methodologies according to which seismically isolated structures can be designed. Both the isolation system and the isolated structure are required to be designed to resist the deformation and stresses produced by seismic events. Two levels of earthquake input are considered in design. The design earthquake (475-year return period) is used to calculate the total design displacement of the isolation system and the lateral forces and displacements of the isolated structure, and the maximum considered earthquake (1000-year return period) is used to calculate the total maximum displacement of the isolation system to ensure its integrity even at extreme ground shaking. Deformation characteristics of the isolation system is requiredto be based on properly substantiated prototype tests with predefined sequence and number of loading cycles.

Seismic isolation is proven to be a very effective method for protecting buildings and other structures against seismic hazards (Skinner et al., 1993; Kitagawa and Midorikawa, 1998; Komodromos, 2001). An important disadvantage of the method, however, is that it cannot be applied partially to structures, unlike most other seismic retrofitting methods. For this reason, the cost of base isolation is often significantly higher than alternative retrofitting methods. This often limits the application of base isolation to

(1) special buildings, such as certain industrial, research, public and hospital buildings that contain sensitive equipment or strict operational and performance requirements,
(2) historical buildings, the architectural and historic character of which may be harmed by alternative retrofitting methods,
(3) bridges, for which relatively less number of isolators are required and installation is easier.

In order to increase the cost competitiveness of base isolation for buildings, there is need for research in the areas of reducing the application costs through efficient design and specialized equipment, and optimization of isolator types, combination, and arrangement.

Supplemental Energy Dissipation and Structural Control An alternative and often more cost efficient retrofitting strategy compared to base isolation is installation of supplemental energy dissipation devices in structures as a means for passive or active structural control. The objective of structural control is to reduce structural vibrations for improved safety and/or serviceability under wind and earthquake loadings. Passive control systems reduce structural vibration and associated forces through energy dissipation devices that do not require external power. These devices utilize the motion of the structure to develop counteracting control forces and absorb a portion of the input seismic energy. Active control systems, however, enhance structural response through control forces developed by force delivery devices that rely on external power to operate. The actuator forces are controlled by real time controllers that process the information obtained from sensors within the structure. Semi-active control systems combine passive and active control devices and are sometimes used to optimize the structural performance with minimal external power requirements.

The severity of seismic demand on a structure is proportional to its stiffness and inversely proportional to its damping or energy dissipation capacity. Thus, installing supplemental energy dissipating devices in the structure reduces the seismic demand and results in increased safety of the structure and its contents from the damaging effects of earthquakes. In recent years, considerable attention has been paid to research and development of structural control devices, with particular emphasis on improving wind and seismic response of buildings and bridges.

In both areas, efforts have been made to develop the structural control concept into a workable technology, and as a result, such devices have been installed in a variety of structures around the world. The most challenging aspect of vibration control research in civil engineering is the fact that this is a field that requires integration of a number of diverse disciplines, some of which

are not within the domain of traditional civil engineering. These include computer science, data processing, control theory, material science, sensing technology, as well as stochastic processes, structural dynamics, and wind and earthquake engineering. These coordinated efforts have facilitated collaborative research among researchers from a diverse background and have accelerated the research to the implementation process. Continued research is essential in this area to develop effective and affordable retrofitting solutions for structures with insufficient seismic resistance.

A special concern regarding the use of energy dissipation devices in structures with high characteristic variability is the fact that the effectiveness of such devices is dependent on the deformation capacity of the structure. For structures that suffer from inadequate seismic detailing, which translates into insufficient deformation capacity, great caution must be exercised in use of these devices for seismic retrofitting. A feasible solution may be to combine this technique with deformation enhancement measures to ensure their effectiveness. This constitutes an important research area with valuable potential contribution and high potential benefits.

Effects of Seismic Retrofitting on Structural Performance

The seismic retrofit techniques briefly presented in the preceding sections vary in the mechanisms that they decrease the seismic risk of structures.

Selection of a particular retrofitting technique depends on the seismic demand, structural capacity, the required performance level, functional characteristics and the importance of the structure. The main challenge is to achieve a desired performance level at a minimum cost, which can best be achieved through a detailed nonlinear analysis.

Soil Structure Interaction

Most of the civil engineering structures involve some type of structural element with direct contact with ground. When the external forces, such as earthquakes, act on these systems, neither the structural displacements nor the ground displacements, are independent of each other. The process in which the response of the soil influences the motion of the structure and the motion of the structure influences the response of the soil is termed as soil-structure interaction (SSI).

Conventional structural design methods neglect the SSI effects. Neglecting SSI is reasonable for light structures in relatively stiff soil such as low rise buildings and simple rigid retaining walls. The effect of SSI, however, becomes prominent for heavy structures resting on relatively soft soils for example nuclear power plants, high rise buildings and elevated-highways on soft soil.

Damage sustained in recent earthquakes, such as the 1995 Kobe Earthquake, have also highlighted that the seismic behaviour of a structure is highly influenced not only by the response of the superstructure, but also by the response of the foundation and the ground as well. Hence, the modern seismic design codes, such as Standard Specifications for Concrete Structures: Seismic Performance Verification JSCE 2005 stipulate that the response analysis should be conducted by taking into consideration a whole structural system including superstructure, foundation and ground.

Effect of Soil Structure Interaction on Structural Response

It has conventionally been considered that soil-structure interaction has beneficial effect on the seismic response of a structure. Many design codes have suggested that the effect of SSI can reasonably be neglected for the seismic analysis of structures. This myth about SSI apparently stems from the false perception that SSI reduces the overall seismic response of a structure, and hence, leads to improved safety margins.

Most of the design codes use oversimplified design spectra, which attain constant acceleration up to a certain period, and thereafter decreases monotonically with period. Considering soil-structure interaction makes a structure more flexible and thus, increasing the natural period of the structure compared to the corresponding rigidly supported structure. Moreover, considering the SSI effect increases the effective damping ratio of the system. The smooth idealization of design spectrum suggests smaller seismic response with the increased natural periods and effective damping ratio due to SSI. With this assumption, it was traditionally been considered that SSI can conveniently be neglected for conservative design. In addition, neglecting SSI tremendously reduces the complication in the analysis of the structures which has tempted designers to neglect the effect of SSI in the analysis.

This conservative simplification is valid for certain class of structures and soil conditions, such as light structures in relatively stiff soil. Unfortunately, the assumption does not always hold true. In fact, the SSI can have a detrimental effect on the structural response, and neglecting SSI in the analysis may lead to unsafe design for both the superstructure and the foundation.

Detrimental Effects of SSI

Using rigorous numerical analyses, Mylonakis and Gazetas have shown that increase in natural period of structure due to SSI is not always beneficial as suggested by the simplified design spectrums. Soft soil sediments can significantly elongate the period of seismic waves and the increase in natural period of structure may lead to the resonance with the long period ground vibration. Additionally, the study showed that ductility demand can significantly increase with the increase in the natural period of the structure due to SSI effect. The permanent deformation and failure of soil may further aggravate the seismic response of the structure.

When a structure is subjected to an earthquake excitation, it interacts with the foundation and the soil, and thus changes the motion of the ground. Soil-structure interaction broadly can be divided into two phenomena: a) kinematic interaction and b) inertial interaction. Earthquake ground motion causes soil displacement known as free-field motion. However, the foundation embedded into the soil will not follow the free field motion. This inability of the foundation to match the free field motion causes the kinematic interaction. On the other hand, the mass of the super-structure transmits the inertial force to the soil causing further deformation in the soil, which is termed as inertial interaction.

At low level of ground shaking, kinematic effect is more dominant causing the lengthening of period and increase in radiation damping. However, with the onset of stronger shaking, near-field soil modulus degradation and soil-pile gapping limit radiation damping, and inertial interaction becomes predominant causing excessive displacements and bending strains concentrated near the ground surface resulting in pile damage near the ground level.

Observations from recent earthquakes have shown that the response of the foundation and soil can greatly influence the overall structural response. There are several cases of severe damages in structures due to SSI in the past earthquakes. Yashinsky cites damage in number of pile-supported bridge structures due to SSI effect in Loma Prieta Earthquake in San Francisco in 1989. Extensive numerical analysis carried out by Mylonakis and Gazetas have attributed SSI as one of the reasons behind the dramatic collapse of Hanshin Expressway in 1995 Kobe Earthquake.

6

The Upper Structure

An upper structure for a bridge includes a coping placed on the top end of a pier, and a girder held by the coping, wherein a side surface of the coping and an end surface of the girder are configured as inclined surfaces (or vertical surfaces), wherein a shear key protrudes on one of the inclined surfaces (or vertical surfaces), and a shear key slot is formed in another one of the inclined surfaces (or vertical surfaces) so as to be engaged with the shear key.

The present disclosure can reduce the construction cost, can realize improved structural efficiency of the bridge upper structure, can realize an easy installation of the girders, can easily combine the girders with the coping without plastering or fixing with mortar by site work, and can efficiently resist to a shear stress that may be generated in the bridge.

Description

Technical Field

The present disclosure relates generally to upper structures for bridges and, more particularly, to an upper structure for bridges in which a coping and girders behave in an integrated state. The present application claims the benefit of Korean Patent Application No. 10-2010-0095326, filed on Sep. 30, 2010, the contents of which are entirely incorporated herein by reference.

Background Art: Generally, bridges are structures that are constructed in various types and various shapes considering the types of objects to be supported by the bridges and to uses of the objects. Further, the bridges function to safely keep up the functions of passageways or facilities supported by the bridges, so the bridges must have a sufficient degree of strength and endurance.

A conventional upper structure for bridges includes a coping 2 that is placed on the top end of a pier 1, girders 4 that are held on the coping 2 with the interposition of respective bridge bearings 3, and a deck (not shown) that is laid on the girders 4 and forms a passageway for vehicles.

Technical Problem

However, in the conventional upper structure for bridges, to realize the continuous arrangement of girders, it is required to necessarily use bridge bearings.

Further, in the conventional upper structure for bridges, the girders and the coping are configured to behave individually, and so the conventional upper structure for bridges is mechanically inefficient. Further, in the conventional upper structure for bridges, the coping is excessively exposed to the outside and spoils the appearance of the bridges.

Accordingly, the present disclosure has been made keeping in mind the above problems occurring in the prior art, and an object of the present disclosure is to provide an upper structure for bridge, in which girders can be installed continuously without using conventional bridge bearings, in which a coping and the girders can behave in an integrated state.

Technical Solution

In order to accomplish the above object, the present disclosure provides an upper structure for a bridge, comprising a coping placed on the top end of a pier, and a girder held by the coping, wherein a side surface of the coping and an end surface of the girder are configured as inclined surfaces so as to come into close contact with each other, wherein a shear key protrudes on one of the inclined surfaces, and a shear key slot is formed in another one of the inclined surfaces so as to be engaged with the shear key.

The coping may include: a horizontal part on which a lower surface of an end of the girder is seated in a state in which the lower surface of the end of the girder comes into close contact with the horizontal part, and a vertical part against which the end surface of the girder abuts in a state in which the end surface of the girder comes into close contact with the vertical part.

The shear key and the shear key slot may comprise a plurality of shear keys that are formed in an inclined surface direction and a plurality of shear key slots that are formed in the inclined surface direction.

The shear key or the shear key slot may be continuously or intermittently formed in a longitudinal direction of the coping (a width direction of the bridge).

In the upper structure for the bridge, an inclined surface directional length of the shear key may be shorter than an inclined surface directional length of the shear key slot, and so the shear key can be easily and closely engaged with the shear key slot.

The shear key may protrude from an associated inclined surface so as to form a trapezoidal shape, and the shear key slot may be depressed in an associated inclined surface so as to form a trapezoidal shape.

The shear key may be formed on the coping and may have a protrusion upper surface that is a horizontal surface, and the shear key slot may be formed in the girder and may have a depression upper surface that is a horizontal surface.

The shear key may be formed in the girder and may have a protrusion lower surface that is a horizontal surface, and the shear key slot may be formed in the coping and may have a depression lower surface that is a horizontal surface.

In another aspect, the present disclosure provides an upper structure for a bridge, comprising a coping placed on a top end of a pier, and a girder held by the coping, wherein a side surface of the coping and an end surface of the girder are configured as vertical surfaces so as to come into close contact with each other, wherein a shear key protrudes on one of the vertical surfaces that come into close contact with each other, and a shear key slot is formed in another one of the vertical surfaces so as to be engaged with the shear key.

The shear key may protrude from an associated vertical surface so as to form a saw tooth shape, and the shear key slot may be depressed in an associated vertical surface so as to form a saw tooth shape.

A plurality of prestressed members may be installed in the coping along a longitudinal direction of the coping so as to prestress the bridge in a width direction of the bridge.

The coping and the girder may be connected to each other by a plurality of prestressed members that are installed along a longitudinal direction of the girder so as to prestress the bridge in a longitudinal direction of the bridge.

Advantageous Effects

As described above, the upper structure for a bridge according to the present disclosure is advantageous in that the girders can be installed continuously without using conventional bridge bearings, thereby reducing the construction cost, in that the coping and the girders behave in an integrated state, thereby realizing improved structural efficiency of the bridge upper structure, and in that the junction surfaces of the coping and the girders are configured as inclined surfaces, thereby realizing an easy installation of the girders on the coping, and in that the shear keys are provided on the inclined surfaces of the coping and the girders, thereby easily combining the girders with the coping without plastering or fixing with mortar by site work, and in that the bridge can efficiently resist to a shear stress that may be generated in the structure of the bridge during the use of the bridge.

Historic Wells Street Bridge in Chicago Pieced Together

The historic Wells Street Bridge is a double-deck, double-leaf, fixed-trunnion bascule bridge. It crosses the Chicago River and carries the double-track Chicago Transit Authority elevated railway on the upper level and vehicle, bicycle and pedestrian traffic on the lower level. The overall length of the main span is 345 ft and the overall width is 72 ft. The Wells Street Bridge was built in 1922 and is listed on the Illinois Historic Bridge Survey.

The bridge superstructure was in poor condition with a current Inventory Rating of HS 4.7 and Operating Rating of HS 7.9, with a posted load limit of 10 tons.

The sufficiency rating for the structure was only 19.0. Reconstruction of this project was one of the top priorities for the Chicago Department of Transportation (CDOT), Division of Engineering, which owns and maintains the bridge. The bridge has been in operation far longer than the 50- to 60-year lifespan of a typical movable bridge. This longevity can be attributed to a combination of robust original construction and routine repairs.

Engineering

Although one of the key elements of this project is the structural design, equally important components of this project are mechanical and electrical engineering, bridge houses, counterweight design, staging to have the least impact to CTA services and coordination with various agencies as the project is located in the heart of downtown Chicago and is surrounded by major traffic corridors and businesses.

Dual truss girders, as main load-carrying members, support both level of framing supported by transverse floor beams spaced about 14 ft. At the lower level, metal grating is supported by longitudinal roadway stringers which span between transverse floor beams. Sidewalks are supported on brackets which are cantilevered from the exterior truss girders. They consist of similar steel grid deck; however, the grid was filled with concrete. At the upper level, the CTA structure is supported by longitudinal track stringers spanning between transverse floor beams. The rear-joint and centre-joint assemblies are critical design elements where the movable and fixed parts come to gather. Two bridge houses are located along the northwest and southeast corners of the movable span.

For the main bridge structure, a major rehabilitation to restore the bridge to an adequate rating required replacement of a number of truss members along with the entire upper- and lower-level framing, floor system and lateral bracings. The "river arm" portion of the trusses from the cantilever tip of the truss to panel points 10-13 were entirely replaced. The remaining portion of the trusses received significant repairs to the bottom chord, verticals and select diagonal members. The centre-break castings and rear brakes needed to be adjusted to accommodate the new floor system. The live-load bearing castings will be cleaned, repaired and reset. Steel components including the trunnion frames, anchor frames and machinery frames also will be repaired. The existing paint has extensive deterioration throughout and the entire bridge will receive a new coat.

Sidewalks are being replaced with fibreglass decking and existing railings will be replaced with new railings to match the original historic railings.

Since significant rehabilitation is to be performed to the structure, a detailed structural analysis was performed to determine the appropriate counterweight adjustments to bring the bridge to a "balanced" condition while keeping track of addition/deletion of weights of the structural components. STAAD-Pro was used to model the bridge superstructure. The upper-level structure supporting the CTA lines was designed using ASD methods in accordance with AREMA and CTA guidelines while the lower-level roadway structure was designed using AASHTO-LRFD specifications.

The bridge is equipped with four machinery drives—two machinery drives for each truss. The prime mover of each drive is a 100-hp 550 VDC mill-type electric motor. These are the original motors that were installed in the 1920s. Each machinery drive is equipped with a machinery brake and a motor brake. The machinery brake is a hydraulic thrustor operated by a wheel and shoe-type brake. The motor brake is a solenoid-operated wheel and shoe-type brake. The motor brake wheel, 22 in. in diameter, is directly connected to the electric motor shaft.

The bridge machinery drive motors will be removed for complete refurbishing. Refurbishing of the original motors was selected over complete motor replacement to retain the same geometric relationship between the motor-shaft centreline and the centreline of mating spur gears as they are mounted upon the existing machinery frame. Refurbishing can be readily performed in a motor rebuilding house and be tested to meet original torque-speed requirements. The project included the replacement of the existing solenoid-operated motor brakes with hydraulically operated motor brakes.

The rear-lock mechanism consists of a pair of motors which drives the speed-reducing and load-transfer mechanism to a toggle linkage locking mechanism that locks into a receiving bracket mounted on the rear of the truss and resists live loads applied between the rear bridge leaf break and the trunnion. These rear locks are driven with a pair of 7.5-hp AC motors with integral motor brakes. These motors are connected to a common shaft which drives a crank shaft, linkage and toggle-type linkage assembly to insert an upper rocker casting beneath the heel lock shoe of the roadside of the bridge truss.

Centre locks are provided at the top chord. There is one linkage on each of the top chords driven by a linkage powered from two 3-hp, 900-rpm motors. The entire centre-lock mechanism will be removed and sent to a machine shop for refurbishing.

The power to operate the bridge is provided by two independent electrical sources. The south side of the bridge is fed from a three-phase, 480-volt AC Commonwealth Edison utility service feed. The north side of the bridge is fed from a 600-volt DC feed provided by the CTA.

The bridge's south side feed was installed during the Lower Wacker construction project around 2000. The three-phase, 480-volt AC power is converted into 600-volt DC power by two rectifier transformers. The DC power is then distributed to the bridge motors via the DC controller and resistor banks. A 480-120/208V transformer was installed to provide power to the south leaf bridge equipment requiring AC power (such as lights, receptacles, gates, etc.) in the south tower as well as AC power to the north side via submarine cable. The CTA feed for the north leaf is being replaced with a three-phase, 480-volt AC feed from ComEd similar to the feed installed for the south leaf of the bridge.

Each bridge house has a controller which controls its associated bridge leaf. New control consoles will be installed. The bridge's up/down direction and speed is controlled by a trolley-car-type control lever and a foot-pedal brake located on the floor of the tower. Equipment such as the trolley control lever, gauges and switches are from the original bridge construction. New submarine cables are being installed to power up the north side DC motor leafs as well as AC power and controls.

The bridge is listed on the National Register of Historic Structures. To that extent, a detailed historical research was conducted to identify critical historical features that must be restored to maintain the original characteristics of the bridge. Historical drawings, photographs and material used since construction of the bridge were reviewed in detail. The project has gone through extensive coordination with the Illinois Historic Preservation Agency (IHPA) to obtain their approval for the proposed repairs/modifications to the bridge structure. The following were the key outcomes of the historical review:

- The new railing should match the original cast-iron ornamental railings;
- New truss elements must match the original open-web, built-up lattice sections;
- Existing truss-type stringers at the upper level can be replaced with solid web-plate girders; and
- Existing rivets may be replaced with high-strength bolts.

The bridge houses will receive select repairs and upgrades to the interiors, roof, walls and windows.

Construction Staging and Coordination with CTA

To ensure a successful completion of the reconstruction of this bridge, an effective communication and coordination between CDOT and CTA was the key. CTA operates its two major service lines over the upper portion of the bascule bridge. The train operations are critical to commuters as well as tourists thus critical for businesses in downtown. During an early stage of the project, coordination and detailed discussions were conducted to determine allowable windows of track shut downs. These shut downs were necessary in order to replace a major portion of trusses and the entire upper-level framing carrying CTA tracks.

Only one bridge leaf was reconstructed at a time. This allowed accommodation of the river traffic below the bridge. The vehicular and pedestrian traffic from the lower level were safely detoured.

Replacement of truss river arms, entire upper-level framing and select lower-level framing to maintain the stability of the trusses was performed in two shutdowns. CTA allowed two windows of nine-day shutdowns, spaced about six weeks apart, for the replacement of the trusses. Construction of the project started in the fall of 2012. River arm trusses were procured and fabricated earlier under a separate contract to expedite the project schedule. The south leaf was replaced during the first nine-day shutdown, followed by the north leaf during the second nine-day shutdown in the spring of 2013.

The first shoring towers were installed under the counterweights and the estimated unbalanced load was jacked into the shores. Initial and interim balance checks were performed, including temporary loading of replaced spans, to maintain the balance. Existing trusses were removed at panel points 10 and 13 and lowered on the barges and moved out of the way. A "catcher beam" along the lower chord of the existing truss was installed to provide stability and partial support during fit up of the new truss. The new truss assembly was completely prefabricated, including all major components of the trusses, off-site and floated on a barge into the final position. New trusses were jacked into the final position using shoring towers on a barge and connected with existing trusses.

Howrah Bridge

The Howrah Bridge, is a cantilever bridge with a suspended span over the Hooghly River in West Bengal, India. Commissioned in 1943, the bridge was originally named the New Howrah Bridge, because it replaced a pontoon bridge at the same location linking the two cities of Howrah and Kolkata (Calcutta). On 14 June 1965 it was renamed Rabindra Setu, after the great Bengali poet Rabindranath Tagore, who was the first Indian and Asian Nobel laureate. It is still popularly known as the Howrah Bridge.

The bridge is one of four on the Hooghly River and is a famous symbol of Kolkata and West Bengal. The other bridges are the Vidyasagar Setu (popularly called the Second Hooghly Bridge), the Vivekananda Setu, and the newly built Nivedita Setu. It weathers the storms of the Bay of Bengal region, carrying a daily traffic of approximately 100,000 vehicles and possibly more than 150,000 pedestrians, easily making it the busiest cantilever bridge in the world. The third-longest cantilever bridge at the time of its construction, the Howrah Bridge is the sixth-longest bridge of its type in the world.

History

1862 proposal by Turnbull: Pontoon bridge

Figure: *The old Pontoon Bridge on the Hooghly River, c. 1901*

In face of the increasing traffic across the Hooghly river, a committee was appointed in 1855-56 to review alternatives for constructing a bridge across it. The plan was shelved in 1859-60, to be revived in 1868, when it was decided that a bridge should be constructed and a newly appointed trust vested to manage it. The Calcutta Port Trust was founded in 1870, and the Legislative department of the then Government of Bengal passed the Howrah Bridge Act in the year 1871 under the Bengal Act IX of 1871, empowering the Lieutenant-Governor to have the bridge constructed with Government capital under the aegis of the Port Commissioners.

Eventually a contract was signed with Sir Bradford Leslie to construct a pontoon bridge. Different parts were constructed in England and shipped to Calcutta, where they were assembled together. The assembling period was fraught with problems. The bridge was considerably damaged by the great cyclone on 20 March 1874. A steamer named *Egeria* broke from her moorings and collided head-on with the bridge, sinking three pontoons and damaging nearly 200 feet of the bridge. The bridge was completed in 1874, at a total cost of 2.2 million, and opened to traffic on 17 October of that year. The bridge was then 1528 ft. long and 62 ft. wide, with 7-foot wide pavements on either side. Initially the bridge was periodically unfastened to allow steamers and other marine vehicles to pass through. Before 1906, the bridge used to be undone for the passage of vessels during daytime only. Since June of that year it started opening at night for all vessels except ocean steamers, which were required to pass through during daytime. From 19 August 1879, the bridge was illuminated by electric lamp-posts, powered by the dynamo at the Mullick Ghat Pumping Station. As the bridge could not handle the rapidly increasing load, the Port Commissioners started planning in 1905 for a new improved bridge.

Plans for a New Bridge

In 1906 the Port Commission appointed a committee headed by R.S. Highet, Chief Engineer, East Indian Railway and W.B. MacCabe, Chief Engineer, Calcutta Corporation. They submitted a report stating that

Bullock carts formed the eight - thirteenths of the vehicular traffic (as observed on 27 August 1906, the heaviest day's traffic observed in the port of Commissioners 16 days' Census of the vehicular traffic across the existing bridge). The road way on the existing bridge is 48 feet wide except at the shore spans where it is only 43 feet in road ways, each 21 feet 6 inches wide. The roadway on the new bridge would be wide enough to take at least two lines of vehicular traffic and one line of trams in each direction and two roadways each 30 feet wide, giving a total width of 60 feet of road way which are quite sufficient for this purpose.................... The traffic across the existing floating bridge Calcutta & Howrah is very heavy and it is obvious if the new bridge is to be on the same site as the existing bridge, then unless a temporary bridge is provided, there will be serious interruptions to the traffic while existing bridge is being moved to one side to allow the new bridge to be erected on the same site as the present bridge.

The committee considered six options:

1. Large ferry steamers capable of carrying vehicular load (set up cost 900,000, annual cost 437,000)
2. A transporters bridge (set up cost 2 million)
3. A tunnel (set up cost 338.2 million, annual maintenance cost 1779,000)
4. A bridge on piers (set up cost 22.5 million)
5. A floating bridge (set up cost 2140,000, annual maintenance cost Rs.200,000)
6. An arched bridge

The committee eventually decided on a floating bridge. It extended tenders to 23 firms for its design and construction. Prize money of £ 3,000 (45,000, at the then exchange rate) was declared for the firm whose design would be accepted.

Planning and Estimation

The initial construction process of the bridge was stalled due to the World War I, although the bridge was partially renewed in 1917 and 1927. In 1921 a committee of engineers named the 'Mukherjee Committee' was formed, headed by Sir R.N. Mukherjee, Sir Clement Hindley, Chairman of Calcutta Port Trus,t and J. McGlashan, Chief Engineer. They referred the matter to Sir Basil Mott, who proposed a single span arch bridge.

In 1922 the New Howrah Bridge Commission was set up, to which the Mukherjee Committee submitted its report. In 1926 the New Howrah Bridge Act passed. In 1930 the Goode Committee was formed, comprising S.W. Goode as President, S.N. Mallick, and W.H. Thompson, to investigate and report on the advisability of constructing a pier bridge between Calcutta and Howrah. Based on their recommendation,

M/s. Rendel, Palmer and Tritton were asked to consider the construction of a suspension bridge of a particular design prepared by their chief draftsman Mr. Walton. On basis of the report, a global tender was floated. The lowest bid came from a German company, but due to increasing political tensions between Germany and Great Britain in 1935, it was not given the contract. The British firm Cleveland Bridge & Engineering Company was awarded the construction contract that year. The New Howrah Bridge Act was amended in 1935 to reflect this, and construction of the bridge started the next year.

Construction

The bridge does not have nuts and bolts, but was formed by riveting the whole structure. It consumed 26,500 tons of steel, out of which 23,000 tons of high-tensile alloy steel, known as Tiscrom, were supplied by Tata Steel. The main tower was constructed with single monolith caissons of dimensions 55.31 x 24.8 m with 21 shafts, each 6.25 metre square. The fabrication was done by Braithwaite, Burn & Jessop Construction Company at four different shops in Kolkata. The two anchorage caissons were each 16.4 m by 8.2 m, with two wells 4.9 m square. The caissons were so designed that the working chambers within the shafts could be temporarily enclosed by steel diaphragms to allow work under compressed air if required. The caisson at Kolkata side was set at 31.41 m and that at Howrah side at 26.53 m below ground level.

One night, during the process of grabbing out the muck to enable the caisson to move, the ground below it yielded, and the entire mass plunged two feet, shaking the ground. The impact of this was so intense that the seismograph at Kidderpore registered it as an earthquake and a Hindu temple on the shore was destroyed, although it was subsequently rebuilt. While muck was being cleared, numerous varieties of objects were brought up, including anchors, grappling irons, cannons, cannon balls, brass vessels, and coins dating back to the East India Company.

The job of sinking the caissons was carried out round-the-clock at a rate of a foot or more per day. The caissons were sunk through soft river deposits to a stiff yellow clay 26.5 m below ground level. The accuracy of sinking the huge caissons was exceptionally precise, within 50–75 mm of the true position. After penetrating 2.1 m into clay, all shafts were plugged with concrete after individual dewatering, with some 5 m of backfilling in adjacent shafts. The main piers on the Howrah side were sunk by open wheel dredging, while those on the Kolkata side required compressed air to counter running sand. The air pressure maintained was about 40 lbs per square inch (2.8 bar), which required about 500 workers to be employed. Whenever excessively soft soil was encountered, the shafts symmetrical to the caisson axes were left unexcavated to allow strict control. In very stiff clays, a large number of the internal wells were completely undercut, allowing the whole weight of the caisson to be carried by the outside skin friction and the bearing under the external wall. Skin friction on the outside of the monolith walls was estimated at 29 kN/m^2 while loads on the cutting edge in clay overlying the founding stratum reached 100 tonnes/m. The work on the foundation was completed on November 1938.

By the end of 1940, the erection of the cantilevered arms was commenced and was completed in mid-summer of 1941. The two halves of the suspended span, each 282 feet (86 m) long and weighing 2,000 tons, were built in December 1941. The bridge was erected by commencing at the two anchor spans and advancing towards the centre, with the use of creeper cranes moving along the upper chord. 16 hydraulic jacks, each of which had an 800-ton capacity, were pressed into service to join the two halves of the suspended span.

The entire project cost 25 million (£2,463,887). The project was a pioneer in bridge construction, particularly in India, but the government did not have a formal opening of the bridge due to fears of attacks by Japanese planes fighting the Allied Powers. Japan had attacked the United States at Pearl Harbor on December 7, 1941. The first vehicle to use the bridge was a solitary tram.

Description

Specifications:

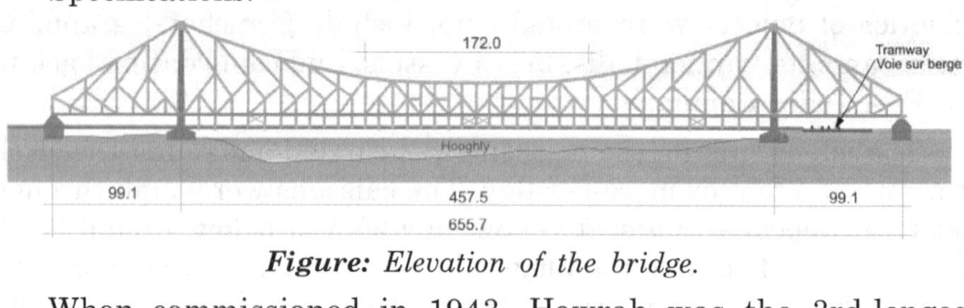

***Figure:** Elevation of the bridge.*

When commissioned in 1943, Howrah was the 3rd-longest cantilever bridge in the world, behind Pont de Québec (549 m) in Canada and Forth Bridge (521 m) in Scotland. It has since been surpassed by three bridges, making it the sixth-longest cantilever bridge in the world in 2013. It is a Suspension type Balanced Cantilever bridge, with a central span 1500 ft between centres of main towers and a suspended span of 564 ft. The main towers are 280 ft high above the monoliths and 76 ft apart at the top. The anchor arms are 325 ft each, while the cantilever arms are 468 ft each. The bridge deck hangs from panel points in the lower chord of the main trusses with 39 pairs of hangers. The roadways beyond the towers are supported from ground, leaving the anchor arms free from deck load. The deck system includes cross girders suspended between the pairs of hangers by a pinned connection. Six rows of longitudinal stringer girders are arranged between cross girders. Floor beams are supported transversally on top of the stringers, while themselves supporting a continuous pressed steel troughing system surfaced with concrete.

The longitudinal expansion and lateral sway movement of the deck are taken care of by expansion and articulation joints. There are two main expansion joints, one at each interface between the suspended span and the cantilever arms, and there are others at the towers and at the interface of the steel and concrete structures at both approach. There are total 8 articulation joints, 3 at each of the cantilever arms and 1 each in the suspended portion. These joints divide the bridge into segments with vertical pin connection between them to facilitate rotational movements of the deck. The bridge deck has longitudinal ruling gradient of 1 in 40 from either end, joined by a vertical curve of radius 4000 ft. The cross gradient of deck is 1 in 48 between kerbs.

Traffic

Traffic Flow for fast moving light vehicles		
Year	***Two-wheelers/Autos***	***Cars/Taxis***
1959	2.47%	97.53%
1986	24%	76%
1990	27%	73%
1992	26%	74%
1999	20%	80%

The bridge serves as the gateway to Kolkata, connecting it to the Howrah Station, which is one of the four intercity train stations serving Howrah and Kolkata. As such, it carries the near entirety of the traffic to and from the station, taking its average daily traffic close to nearly 1.5 million pedestrians and 1 million vehicles. In 1946 a census was taken to take a count of the daily traffic, it amounted to 27,400 vehicles, 121,100 pedestrians and 2,997 cattle. The bulk of the vehicular traffic comes from buses and cars. Prior to 1993 the bridge used to carry trams also. Trams departed from the terminus at Howrah station towards Rajabazar, Sealdah, High Court, Dalhousie Square, Park Circus and Shyambazar.

From 1993 the tram services on the bridge were discontinued due to increasing load on the bridge. However the bridge still continues to carry much more than the expected load. A 2007 report revealed that nearly 90,000 vehicles were plying on the bridge daily (15,000 of which were goods-carrying), though its load-bearing capacity is only 60,000. One of the main reasons of overloading was that although vehicles carrying up to 15 tonnes are allowed on the structure, vehicles with 12-18 wheels and carrying load up to 25 tonnes often plied on it. 31 May 2007 onwards, overload trucks were banned from plying on the bridge, and were redirected to the Vidyasagar Setu instead. The road is flanked by footpaths of width 15 feet, and they swarm with pedestrians.

Maintenance

The Kolkata Port Trust is vested with the maintenance of the bridge. The bridge has been subject to damage from vehicles due to rash driving, and corrosion due to atmospheric conditions and biological wastes. On October 2008, 6 high-tech surveillance cameras were placed to monitor the entire 705-metre-long and 30-metre-wide structure from the control room. Two of the cameras were placed under the floor of the bridge to track the movement of barges, steamers and boats on the river, while the other four were fixed to the first layer of beams — one at each end and two in the middle — to monitor vehicle movements. This was in response to substantial damage caused to the bridge from collisions with vehicles, so that compensation could be claimed from the miscreants.

Corrosion has been caused by bird droppings and human spitting. An investigation in 2003 revealed that as a result of prolonged chemical reaction caused by continuous collection of bird excreta, several joints and parts of the bridge were damaged. As an immediate measure, the Kolkata Port Trust engaged contractors to regularly clean the bird droppings, at an annual expense of 500,000.

In 2004, KPT spent 6.5 million to paint the entirety of 2.2 million sq. m of the bridge. Two coats of Aluminium paint, with a primer of

Zinc chromate before that, was applied on the bridge, requiring a total of 26,500 litres of paint.

The bridge is also considerably damaged by human spitting. A technical inspection by Port Trust officials in 2011 revealed that spitting had reduced the thickness of the steel hoods protecting the pillars from six to less than three millimeters since 2007. The hangers need those hoods at the base to prevent water seeping into the junction of the cross-girders and hangers, and damage to the hoods can jeopardize the safety of the bridge. KPT announced that it will spend 2 million on covering the base of the steel pillars with fibreglass casing to prevent spit from corroding them.

On 24 June 2005, a private cargo vessel M V Mani, belonging to the Ganges Water Transport Pvt. Ltd, while trying to pass under the bridge during high tide, had its funnel stuck underneath for three hours, causing substantial damage worth about 15 million to the stringer and longitudinal girder of the bridge. Some of the 40 cross-girders were also broken. Two of four trolley guides, bolted and welded with the girders, were extensively damaged. Nearly 350 of 700 metres of the track were twisted beyond repair. The damage was so severe that KPT requested help from Rendall-Palmer & Tritton Limited, the original consultant on the bridge from UK. KPT also contacted SAIL to provide 'matching steel' used during its construction in 1943, for the repairs. For the repair costing around 5 million, about 8 tonnes of steel was used. The repairs were completed in early 2006.

Overpass

An overpass (called a flyover in the United Kingdom and most Commonwealth countries) is a bridge, road, railway or similar structure that crosses over another road or railway. An *overpass* and *underpass* together form a grade separation. Stack interchanges are made up of many overpasses.

History

The world's first railroad flyover was constructed in 1843 by the London and Croydon Railway at Norwood Junction railway station to carry its atmospheric railway vehicles over the Brighton Main Line.

Highway and Road

In North American usage, a *flyover* is a high-level overpass, built above main overpass lanes, or a bridge built over what had been an at-grade intersection. Traffic engineers usually refer to the latter as a *grade separation.* A flyover may also be an extra ramp added to an existing interchange, either replacing an existing cloverleaf loop (or being built in place of one) with a higher, faster ramp that eventually bears left, but may be built as a right or left exit.

A cloverleaf or partial cloverleaf contains some 270 degree loops, which can slow traffic and can be difficult to construct with multiple lanes. Where all such turns are replaced with flyovers (perhaps with some underpasses) only 90 degree turns are needed, and there may be four or more distinct levels of traffic. Depending upon design, traffic may flow in all directions at or near open road speeds (when not congested).

High Five Interchange

The High Five Interchange is the first five-level stack interchange built in Dallas, Texas. Located at the junction of the Lyndon B. Johnson Freeway (Interstate 635, I-635) and the Central Expressway (US Highway 75, US 75), it replaces an antiquated partial cloverleaf interchange constructed in the 1960s.

The $261 million project was initiated in 2002 and completed in December 2005. It was designed by the HNTB Corporation and built by Zachry Construction Corporation.

The interchange is considered by *Popular Mechanics* to be one of "The World's 18 Strangest Roadways" because of its height (as high as a 12-story building), its 37 permanent bridges and other unusual design and construction features. In 2006, the American Public Works Association named the High Five Interchange as "Public Works Projects of the Year."

Description

The High Five Interchange, north of downtown in Dallas, Texas, is a massive five-level freeway interchange. It is the junction of two major highways carrying heavy rush hour traffic, the Lyndon B. Johnson Freeway (Interstate 635) and the Central Expressway (US 75), and is the first five-level stack interchange to be built in the city.

It replaces the antiquated three-level, modified cloverleaf interchange built in the 1960s that caused a severe bottleneck by narrowing US 75 down to two lanes at the junction of the two highways. The looped ramps of the cloverleaf forced motorists to slow down drastically, backing up traffic. Left-hand exits contributed to the congestion. Further, its two frontage roads were not directly connected to each other, making local access difficult.

Alluding to the celebratory gesture, the "High Five" name refers to the five interchange ramps that tower over the landscape. The interchange is as high as a 12-story building, and includes 37 bridges spread across five levels (the "High Five"), 710 support tiers, and 60 miles of additional highway. The highest ramps are 120 feet (37 m) above ground. The lanes of US 75, which are on the bottom level, are 20 feet (6 m) below ground level. As part of the project, I-635 was widened to include four dedicated high-occupancy vehicle lanes (HOV lanes) that are barrier-separated and reversible.

The roads on the five levels are:

- Level I: US 75, an eight lane highway, four lanes going each way
- Level II: The junction of the two six-lane frontage roads, each having three through lanes in each direction, left-turn lanes and turnarounds, and easy access to US 75 and I-635
- Level III: I-635, ten regular lanes, five going each way, and four HOV lanes (two going each way) separated by barriers
- Level IV and V: Direct connection ramps (two levels), eliminating the left exits of the modified cloverleaf

The interchange, with its decorative etchings on precast concrete elements, along with the visually appealing coloration specified by the Texas Department of Transportation (TxDOT), can be regarded as an enormous work of public art. The High Five also incorporates a hiking and bike trail, named the Cottonwood Trail, which runs under all levels of the interchange. The section of the trail passing beneath the interchange was constructed as part of the High Five project by TxDOT.

The interchange is considered by *Popular Mechanics* as one of "The World's 18 Strangest Roadways", calling it a "labyrinth of lanes" and a "five-level marvel of engineering" because of its height, the number of its bridges and other unusual design and construction features.

History

The High Five Interchange project was planned as a replacement for the existing, antiquated interchange that accommodated 500,000 vehicles daily and was located in one of most intensely developed commercial zones in Dallas. It was a collaborative project between the TxDOT, affected motorists and property owners, and the primary contractor, Zachry Construction. An essential consideration was to complete the project with as little disruption to the traffic flow as possible.

Figure: *High Five Interchange: US 75 on the bottom level; the two frontage roads on the second; I-635 on the third; direct connection ramps on the two top levels*

Reasons for upgrading the interchange, in addition to the need to increase traffic flow and reduce congestion, were to improve air quality, safety, and local access.

Zachry Construction Corporation submitted the lowest bid and was awarded the $261 million contract by TxDOT, the largest ever by that agency. The construction contract for the High Five Interchange was unique in that it contained elements not found in other construction project contracts. For example, it specified the concrete maturity method to be used to ensure the concrete's strength (the first time

TxDOT had ever done so), and contained an early completion bonus, a sliding scale of up to $11 million if the company completed the contract within four years. Also built into the construction contract was a provision that Zachry was charged for "lane rentals" by TxDOT for time they closed down traffic lanes—fees based on hourly assessments and the time of day the lane was closed.

The fees ranged from $50 to a high of $110,000 for rush hours. In order not to interfere with traffic flow during construction, the contractor built flyover ramps over the existing junction which were then removed upon completion. Little of the original interchange remained when the project was finished.

To save time and money, innovative construction methods were employed: The original plans called for the segments of the long ramps and spans (used to direct-connect roads) to be made of steel, but because building with steel would interfere with heavy traffic flow during construction, cast-in-place segmental concrete was substituted in the plans. However, Zachry decided to use precast concrete segmental bridges, rather than casting the bridge elements in place, and used a unique machine, designed and constructed by Deal S.R.L. of Italy and costing about $1 million, to move the pieces into place.

During the construction, 2.2 million cubic yards of earthwork was used, 350,000 cubic yards of concrete was mixed on site, and 300,000 square feet of retaining walls were built. Additionally, 40,000 of drilled shafts and 75,000 linear feet of drainage pipe were completed. In addition to the construction of 37 permanent bridges, six temporary bridges were built, resulting in 2.3 million square feet of bridge deck.

The project was designed by the HNTB Corporation who provided professional engineering consultation throughout the construction, which had begun in 2002 and was completed in December 2005, more than 13 months sooner than planned.

Awards

In 2006, the American Public Works Association named the High Five "Public Works Project of the Year" for its massive size, its innovative design, the complexity and rapidity of its construction and the need it fulfilled for the community. TxDOT as the managing agency, Zachry Construction Corporation as the primary contractor, and HNTB Corporation as the primary consultant received the award in recognition of their cooperative alliance in completing the project.

Pedestrian

A pedestrian overpass allows pedestrians safe crossing over busy roads without impacting traffic.

A pedestrian separation structure is any structure that removes pedestrians from a vehicle roadway. This creates a road junction where vehicles and pedestrians do not interact.

This can be considered a type of grade separation structure on the road.

These structures can be located either above the roadway or below the roadway. In the U.S., access under the Americans with Disabilities Act requirements means that stairs cannot be the only access to these structures. An elevator must be provided or a ramp built that conforms to the grade requirements under the ADA regulations.

In the broadest sense, building codes that limit the number of driveways that cross sidewalks may be viewed as making the sidewalks a separation structure.

In many areas, wildlife crossings are provided in wilderness areas to allow wildlife to cross roadways without risking accidents. While not specifically built for people, they could be used by people in those areas.

Railway

Railway overpasses are used to replace level crossings (at-grade crossings) as a safer alternative. Using overpasses allows for unobstructed rail traffic to flow without conflicting with vehicular and pedestrian traffic. Rapid transit systems use complete grade separation of their rights of way to avoid traffic interference with frequent and reliable service.

Railroads also use balloon loops and flying junctions instead of flat junctions, as a way to reverse direction and to avoid trains conflicting with those on other tracks.

Seismic Design Practice in Japan

Japan is one of the most seismically disastrous countries in the world and has often suffered significant damage from large earthquakes. More than 3000 highway bridges have suffered damage since the 1923 Kanto earthquake. The earthquake disaster prevention technology for highway bridges has been developed based on such bitter damage experiences. Various provisions for designing bridges have been developed to prevent damage due to the instability of soils such as soil liquefaction.

Furthermore, design detailings including unseating prevention devices are implemented. With progress in improving seismic design provisions, damage to highway bridges caused by the earth-quakes has been decreasing in recent years.

However, the Hyogo-ken Nanbu earthquake of January 17, 1995 caused destructive damage to highway bridges. Collapse and near collapse of superstructures occurred at nine sites, and other destructive damage occurred at 16 sites.The earthquake revealed that there are a number of critical issues to be revised in the seismic design and seismic retrofit of bridges.

The damage caused by the 1995 Hyogo-ken Nanbu earthquake, the lessons learned from the earthquake, and the seismic design methods introduced in the 1996 Seismic Design Specifications for Highway Bridges are then described. Seismic performance levels and design methods as well as ductility design methods for reinforced concrete piers, steel piers, foundations, and bearings are described. Then the history of the past seismic retrofit practices is described. The seismic retrofit programme after the Hyogo-ken-Nanbu earthquake is described with emphasis on the seismic retrofit of reinforced concrete piers as well as research and development on the seismic retrofit of existing highway bridge.

1996 Seismic Design Specifications of Highway Bridges

Basic Principles of Seismic Design: The 1995 Hyogo-ken Nanbu earthquake, the first earthquake to be considered that such destructive damage could be prevented due to the progress of construction technology in recent years, provided a largo impact on the carthquake disaster prevention measures in various fields. Part V: Seismic Design of the Design Specifications of Highway Bridges (Japan Road Association) was totally revised in 1996, and the design procedure moved from the traditional seismic coefficient method to the ductility design method. The revision was so comprehensive that the past revisions of the last 30 years look minor.

A major revision of the 1996 Specifications is the introduction of explicit two-level seismic design consisting of the seismic coefficient method and the ductility design method. Because Type I and Type II ground motions are considered in the ductility design method, three design seismic forces are used in design. Seismic performance for each design force is clearly defined in the Specifications.

The bridges are categorized into two groups depending on their importance: standard bridges (Type A bridges) and important bridges (Type B bridges). The seismic performance level depends on the importance of the bridge. For moderate ground motions induced in earthquakes with a high probability of occurrence, both A and B bridges should behave in an elastic manner without essential structural damage. For extreme ground motions induced in earthquakes with a low probability of occurrence, Type A bridges should prevent critical failure, whereas Type B bridges should perform with limited damage.

In the ductility design method, two types of ground motions must be considered. The first is the ground motions that could be induced in plate boundary-type earthquakes with a magnitude of about 8. The ground motion at Tokyo in the 1923 Kanto earthquake is a typical target of this type of ground motion. The second is the ground motion developed in earthquakes with magnitude of about 7 to 7.2 at very short distance. Obviously, the ground motions at Kobe in the Hyogo-ken Nanbu earthquake is a typical target of this type of ground motion. The first and the second ground motions are called Type I and Type II ground motions, respectively. The recurrence time of Type II ground motion may be longer than that of Type I ground motion, although the estimation is very difficult.

The fact that lack of near-field strong motion records prevented serious evaluation of the validity of recent seismic design codes is important. The Hyogo-ken Nanbu earthquake revealed that the history of strong motion recording is very short, and that no near-field records have yet been measured by an earthquake with a magnitude on the order of 8. It is therefore essential to have sufficient redundancy and ductility in a total bridge system.

Design Methods

Japan has twenty years of experience in designing and constructing base-isolated building structures. Construction has increased significantly since the 1995 Hyogoken-Nanbu (Kobe) earthquake, having reached over 150 annual construction projects. Many new developments and refinements have been made in the material, device, design, and construction of these structures. It is notable that the size, height, and fundamental natural period of new base-isolated buildings increase steadily with time, indicating that base-isolation in Japan is reaching maturity. Base-isolators and dampers commonly adopted in Japan are also introduced, with emphasis on recent design efforts to enlarge the natural period of base-isolated structures and reduce the lateral forces induced in the superstructure. Basic design procedures are presented, including determination of design earthquake forces, modelling of base-isolation layers, modelling of the superstructure, selection of ground motions, time-history analyses, and performance criteria.

Bridges are designed by both the seismic coefficient method and the ductility design method.

In the seismic coefficient method, a lateral force coefficient ranging from 0.2 to 0.3 has been used based on the allowable stress design approach. No change has been made since the 1990 Specifications in the seismic coefficient method.

Base Isolation

In recent years base isolation has become an increasingly applied structural design technique for buildings and bridges in highly seismic areas. Many types of structures have been built using this approach, and many others are in the design phase or under construction. Most of the completed buildings and those under construction use rubber isolation bearings in some way in the isolation system.

The ideas behind the concept of base isolation are quite simple. There are two basic types of isolation systems. The system that has been adopted most widely in recent years is typified by the use of elastomeric bearings, the elastomer made of either natural rubber or neoprene. In this approach, the building or structure is decoupled from the horizontal components of the earthquake ground motion by interposing a layer with low horizontal stiffness between the structure and the foundation. This layer gives the structure a fundamental frequency that is much lower than its fixed-base frequency and also much lower than the predominant frequencies of the ground motion. The first dynamic mode of the isolated structure involves deformation only in the isolation system, the structure above being to all intents and purposes rigid. The higher modes that will produce deformation in the structure are orthogonal to the first mode and consequently also to the ground motion. These higher modes do not participate in the motion, so that if there is high energy in the ground motion at these higher frequencies, this energy cannot be transmitted into the structure. The isolation system does not absorb the earthquake energy, but rather deflects it through the dynamics of the system. This type of isolation works when the system is linear and even when undamped; however, some damping is beneficial to suppress any possible resonance at the isolation frequency.

The second basic type of isolation system is typified by the sliding system. This works by limiting the transfer of shear across the isolation interface. Many sliding systems have been proposed and some have been used. In China there are at least three buildings on sliding systems that use a specially selected sand at the sliding interface. A type of isolation containing a lead-bronze plate sliding on stainless steel with an elastomeric bearing has been used for a nuclear power plant in South Africa. The friction-pendulum system is a sliding system using a special interfacial material sliding on stainless steel and has been used for several projects in the United States, both new and retrofit construction.

Research at EERC

Research on the development of natural rubber bearings for isolating buildings from earthquakes began in 1976 at the Earthquake Engineering Research Centre (EERC) (now PEER, the Pacific Engineering Research Centre) of the University of California at Berkeley. The initial research programme was a joint effort by EERC and the Malaysian Rubber Producers Research Association (MRPRA), U.K. The programme was funded by MRPRA through a number of grants over several years, with later funding provided by the National Science Foundation and the Electric Power Research Institute. Professor James M. Kelly directed the research at EERC, which included considerable theoretical and experimental contributions by graduate students.

Although not an entirely new idea at the time—a few methods using rollers or sliders had been proposed—the concept of base isolation was considered to be very impractical by most of the structural engineering profession. The research project began with a set of hand-made bearings of extremely low-modulus rubber used with a simple three-story, single-bay, 20-ton model. Shaking table tests showed that isolation bearings could bring about reductions in acceleration by factors of as much as ten when compared to those of conventional design and that, as predicted, the model would respond as a rigid body with all deformation concentrated in the isolation system. It was also clear that a certain degree of damping was needed in the system and that the scale of the model was too small to allow more practical rubber compounds to be used.

In 1978, a more convincing demonstration of the isolation concept was achieved with a more realistic five-story, three-bay model weighing 40 tons and by using damping-enhanced bearings made by commercial techniques. A strong interest throughout the EERC research programme was in the influence of isolation on the response of equipment and contents in a structure, which tend to sustain more damage when conventional methods of seismic-resistant design are used and which, in many buildings, are much more costly than the structure itself.

An extensive series of tests on the five-story frame demonstrated that isolation with rubber bearings could provide very substantial reductions in the accelerations experienced by internal equipment, exceeding the reductions experienced by the structure. However, the same tests showed that when additional elements (such as steel energy-absorbing devices, frictional systems, or lead plugs in the bearings) were added to the isolation system to increase damping, the reductions in acceleration to the equipment were not achieved because the added elements also induced responses in the higher modes of the structure, affecting the equipment. It became clear that the optimum method of increasing damping was to provide it in the rubber compound itself. This method was applied later in the compound developed by MRPRA and used in the first base-isolated building in the United States, described below.

Rubber bearings are relatively easy to manufacture, have no moving parts, are unaffected by time, and are very resistant to environmental degradation.

Test of bearing used in the Indonesian demonstration building. Photo: I. D. Aiken

The bearings are made by vulcanization bonding of sheets of rubber to thin steel reinforcing plates. Because the bearings are very stiff in the vertical direction and very flexible in the horizontal direction, under seismic loading the bearing layer isolates the building from the horizontal components of the ground movement while the vertical components are transmitted to the structure relatively unchanged. Although vertical accelerations do not affect most buildings, the bearings also isolate the building from unwanted high-frequency vertical vibrations produced by underground railways and local traffic.

Rubber bearings are suitable for stiff buildings up to seven stories in height. For this type of building, uplift on the bearings will not occur and wind load will be unimportant.

U.S. Applications

The first base-isolated building in the United States is the Foothill Communities Law and Justice Centre, a $30 million legal services centre in Rancho Cucamonga San Bernardino County, about 97 km (60 miles) east of downtown Los Angeles. Completed in 1985, the building is four stories high with a full basement and sub-basement for the isolation system, which consists of 98 isolators of multilayered natural rubber bearings reinforced with steel plates. The superstructure of the building has a structural steel frame stiffened by braced frames in some bays.

Foothill Communities Law and Justice Centre

The building is located 20 km (12 miles) from the San Andreas fault. San Bernardino County, the first in the U.S. to have a thorough earthquake preparedness programme, asked that the building be designed for a Richter magnitude 8.3 earthquake, the maximum credible earthquake for that site.

The design selected for the isolation system, which accounted for possible torsion, incorporated a maximum horizontal displacement demand of 380 mm (15 in.) in the isolators at the corners of the building. Tests of full-scale sample bearings verified this capacity.

The highly filled natural rubber from which the isolators are made, developed as part of the EERC research programme, has mechanical properties that make it ideal for a base isolation system. The shear stiffness of this rubber is high for small strains but decreases by a factor of about four or five as the strain increases, reaching a minimum value at a shear strain of 50 percent.

For strains greater than 100 percent, the stiffness begins to increase again, providing a fail-safe action under a very high load. The damping follows the same pattern but less dramatically, decreasing from an initial value of 20 percent to a minimum of 10 percent and then increasing again.

The design of the system assumes minimum values of stiffness and damping and a linear response. The high initial stiffness is invoked only for wind load design and the large strain response only for fail-safe action.

This high-damping rubber system was also adopted for the Fire Department Command and Control Facility (FCCF) of Los Angeles County, completed in 1990. (The same type of high-damping rubber bearing was also used for the Italian telephone company, S.I.P., Ancona, Italy, the first modern base-isolated building in Europe.) The FCCF building houses the computer systems for the emergency services of the county and is therefore required to remain functional after an extreme event.

Fire Department Command and Control Facility

The decision to use base isolation for this project was reached by comparing conventional and isolation schemes designed to provide the same degree of protection. In most projects, the isolation design costs five percent more. Not only was the isolation design estimate 6 percent less in this case but is less for any building when equivalent levels of protection are considered. Furthermore, these costs are first costs. Life-cycle costs are even more favourable. Also noteworthy is that the conventional code design requires only a minimal level of protection, that the structure not collapse; whereas isolation design provides a higher level of protection.

The University of Southern California Teaching Hospital in eastern Los Angeles is an eight-story concentrically braced steel frame supported on 68 lead rubber isolators and 81 elastomeric isolators. The building was instrumented by the California Strong Motion Instrumentation Programme soon after its completion in 1991.

The foundation system consists of spread footings and grade beams on rock. Because of functional requirements, both the building plan and elevation are highly irregular with numerous setbacks over the height. Two wings at either side of the building are connected through what is referred to as the "necked-down" portion of the building, and in the original fixed-base design the irregular configuration led to both coupling between the lateral and torsional vibration modes and very large shear force demands in the slender region between the two rings. (Even in the isolated design steel trusses are required to carry the shears in the necked-down region.) These were two of the main reasons that seismic isolation was eventually chosen for this structure.

University of Southern California University Hospital

The University of Southern California (USC) Teaching hospital was 36 km (23 miles) from the epicenter of the M_w 6.8 1994 Northridge earthquake. The peak ground acceleration outside the building was 0.49 g, and the accelerations inside the building were around 0.10 to 0.13 g. In this earthquake the structure was effectively isolated from ground motions strong enough to cause significant damage to other buildings in the medical centre.

The records obtained from the USC hospital are particularly encouraging in that they represent the most severe test of an isolated building to date.

Nuclear Applications

Isolation used in conventional nuclear plants greatly simplifies the expensive and time-consuming design and qualification of the equipment, piping, and supports for seismic loading. In addition, when seismic design criteria are increased due to the discovery of nearby faults, for example, the plant need not be redesigned; upgrading the isolation system is sufficient.

In an experimental programme at EERC isolation bearings were designed, produced, and tested for two types of liquid metal reactor designs. The first, called PRISM, uses high-shape factor isolation bearings designed to provide horizontal isolation only. In the other design, SAFR, the reactor is supported on low-shape bearings that provide both horizontal and vertical isolation. The results of this test series extended the range of the isolator types with well-understood characteristics.

Base Isolation in Japan

After a slow start, base isolation research and development in Japan increased rapidly. The first large base-isolated building was completed in 1986. Although such buildings in Japan require special approval from the Ministry of Construction, as of June 30, 1998, 550 base-isolated buildings had been approved.

Base isolation has advanced rapidly in Japan for several reasons. The expenditure for research and development in engineering is high with a significant amount designated specifically for base isolation; the large construction companies aggressively market the technology; the approval process for constructing a base-isolated building is a straightforward and standardized process; and the high seismicity of Japan encourages the Japanese to favour the long-term benefits of life safety and building life-cycle costs when making seismic design decisions.

The system most commonly used in the past has been natural rubber bearings with mechanical dampers or lead-rubber bearings. Recently, however, there has been an increasing use of high-damping natural rubber isolators. There are now several large buildings that use these high-damping bearings: an outstanding example is the computer centre for the Tohoku Electric Power Company in Sendai, Miyako Province.

Tohoku Electric Power Company, Japan

Currently the largest base-isolated building in the world is the West Japan Postal Computer Centre, located in Sanda, Kobe Prefecture. This six-story, 47,000 m square (500,000 ft square) structure is supported on 120 elastomeric isolators with a number of additional steel and lead dampers. The building, which has an isolated period of 3.9 sec, is located approximately 30 km (19 miles) from the epicenter of the 1995 Hyogoken Nanbu (Kobe) earthquake, and experienced severe ground motion. The peak ground acceleration under the isolators was 400 cm/sec square (0.41 g) but was reduced by the isolation system to 127 cm/sec square (0.13 g) at the sixth floor. The estimate of the displacement of the isolators is around 12 cm (4.8 in.). A fixed-base building adjacent to the computer centre experienced some damage, but there was no damage to the isolated building.

The use of isolation in Japan continues to increase, especially in the aftermath of the Kobe earthquake. As a result of superior performance of the West Japan Postal Computer Centre, there has been a rapid increase in the number of permits for base-isolated buildings, including many apartments and condominiums.

Akashi-Kaikyo Bridge

Akashi-Kaikyo Bridge is the longest span bridge in the world, located in Akashi Strait between Kobe and Awaji Island in Hyogo Pref. The total and central span lengths are 3,911m and 1991m respectively.

The Bridge was constructed as one of Honshu-Shikoku Bridge project, which is one of the greatest projects in Japan to connect Honshu with Shikoku by both highway and railway. The construction work was started in May, 1988 and completed in April, 1998 after 10 years' construction period.

Akashi Strait ("Kaikyo" means "strait" in English.) is located between Osaka Bay and Seto Inland Sea. The width of the strait is about 4km and its maximum depth is about 110m. The utmost tidal current is about 4.5m/sec (about 9 knots).

The surrounding around Akashi Strait is a well-known fishery area and the Strait is designated as an international channel by the Maritime Traffic Safety Act in Japan. About 1400 ships daily pass the navigation. The strait is one of important points at sea.

Akashi-Kaikyo Bridge was constructed under such severe natural and social conditions.

The Bridge was designed to resist both strong wind, 80m/sec, and great earthquakes such as an inter-plate type earthquake which predicts to occur around Pacific Ocean and an earthquake directly above its epicenter like the Southern Hyogo Prefecture Earthquake in 1995.

In addition, huge foundations had to be constructed in the deep sea of 60m and under the fast tidal current of 4.5m/sec. The structure also has to support the vertical force of 120,000t.

Although the central span length of the Bridge is about twice as long as those of previous suspension bridges, the girder of 1,991m long had to be erected. Even under such circumstances, the Bridge was constructed by adopting Japanese own techniques and newly developed technologies.

The great earthquake of Magnitude 7.2 in Richter scale occurred around Akashi Strait, whose epicenter was 10km to 20km deep, at 5:46 on January 17, 1995.

There were a lot of damages during the Earthquake around Kobe and Awaji Island.

At that time, Akashi-Kaikyo Bridge was under construction, in other words, both the towers were completed and main cables of the Bridge were being built.

After the Earthquake, comprehensive investigation was done. As a result of the survey, it was observed that the span lengths were changed by the Earthquake.

However, it was confirmed that the foundations of the Bridge were moved with the bearing ground without sliding between the foundations and the ground.

After several analyses considering such conditions, it was clear that the influence of the moving was minimal and the construction work was able to continue as it is.

The girder length had to be revised because the length between the towers was changed.

This was solved by changing the girder length which was not fabricated.

Although the severest earthquake hit the Bridge, Akashi-Kaikyo Bridge was completed after 10-year construction period.

Inspection, maintenance and operation have to be given to keep the Bridge in good conditions for more than 200 years from now.

In order to do appropriate management, new technology for the maintenance has been researched and developed. One of them is “Dry air injection system”.

“Dry air injection system” is to protect main cables from corrosion. The system is to inject dry air into main cables and to keep the humidity inside the cables constant. This system is revolutionary idea because the maintenance cost can be saved.

Bridge Construction System

Passing Down the Secrets of Technology to Only One Successor

During the Edo Period (1603 - 1868), important technologies were primarily handed down from the father to the eldest son, or sometimes from the master to his first disciple. This system is indicated by old documents that list the chief carpenters' names. During the Edo Period, the Kintaikyo Bridge projects were implemented by the Iwakuni Domain. Carpenters engaged in the projects were vassals of the lord of Iwakuni. Among the vassals, those who belonged to a specific family succeeded to the role of bridge construction. Accordingly, the bridge building technology was inherited in that family. In handing down technologies, it is essential that successors have first-hand experience. Regarding the technologies involved in building wooden structures, there are aspects that even the best drawing or book cannot record. Without experiencing actual bridge construction, no carpenter can build a wooden bridge like the Kintaikyo.

Historic Materials

Many historic documents remain regarding the Kintaikyo Bridge. Of them, the oldest extant material is the drawing prepared in 1699. (There are 12 additional drawings that remain intact today.) The oldest drawing indicates in detail the types, dimensions and production centres of the timbers used for the Bridge; types and number of necessary metal fittings; primary dimensions and gradients of beams used for the three arches in the Bridge's central portion etc. Without these documents, it would have been impossible to rebuild the Bridge each time it was carried away.

To long sustain the bridge building technology, we must also prepare and preserve such records for future generations.

To Learn from Wood to Know Wood

Since the Edo Period, carpenters have passed down verbally the secrets of their technology, or aspects that are difficult to express either in text or drawing. Among such secrets is the method of learning from wood the nature inherent in each respective wood type.

Each timber has a different and unique nature, which can be learned only by viewing and touching the timber, and using all five senses. Without knowing the nature of each timber, carpenters cannot use it for the appropriate purpose.

Unfortunately, the wide spread of laminated lumber provides contemporary carpenters with few opportunities to learn from wood. In this environment, rebuilding the Kintaikyo Bridge, which employs authentic timbers, is a golden opportunity for carpenters to acquire the skill of learning from wood the nature inherent in each type of wood.

The Order System Prioritizing Technological Transfer

Since the construction of the first bridge, the Kintaikyo has been rebuilt and maintained by local people. To long sustain the traditional bridge building technology, it is best that such technology be passed down to local people.

With this view, in selecting builders the local authority adopts a single tendering method rather than a competitive tender, which is the most common way of ordering public works. Although use of the single tendering method is rare, in the case of the Kintaikyo projects it is most effective in transferring traditional technology.

A drawback of the single tendering method, however, is that project expenses can become higher due to the lack of competitive bidding. Even so, this system is essential for passing down traditional technologies to future generations in the Iwakuni region, as well as for maintaining local carpenters' enthusiasm for inheriting their predecessors' high technical expertise.

Order Placement by Type of Work

Rebuilding the Kintaikyo Bridge requires various types of work, including temporary structure construction (building footings etc.), civil engineering (preparing work yard), carpentry (timber processing and bridge building), metal processing (processing steel and copper plates), painting (painting timbers and steel members for antiseptic treatment), procurement (purchase of timber) and nail production (producing traditional Japanese nails).

Since from the standpoint of managerial efficiency it is unwise to place separate orders for individual types of work, in the recent project the local authority placed orders - orders for temporary structure construction, civil engineering, carpentry, metal processing and painting - collectively with a local builders' union. However, the authority placed orders with timber merchants and nail producers on an individual basis. Instead of supplying raw materials to individual worker unions, the authority supplied them to the principal contractor, i.e. the local builders' union.

As to carpentry work, the primary contractor subcontracted the work to a local carpenters' union. Since carpenters usually work by themselves, the primary contractor subcontracted the project, involving hundreds of millions of yen, to the carpenters' union rather than to individual carpenters. Similarly, the primary contractor subcontracted other types of work to selected local unions, rather than to individual workers. Since this method has been effective in passing down traditional techniques to younger generations, the authority will continue this order placement system in the future.

Fostering Carpenters

Only carpenters who have received sufficient training can process timbers and build a bridge like the Kintaikyo Bridge. To build a structure using traditional technology, carpenters must have expertise in processing timbers, as well as skills in identifying the inherent nature of individual timbers.

Recently, however, carpenters primarily use precut, laminated timbers. Contemporary carpenters therefore have fewer opportunities to use and process authentic, live timbers. Therefore, rebuilding the Kintaikyo Bridge affords carpenters a rare opportunity to learn the traditional technology of wood building. In the rebuilding project conducted during the Heisei Period (1989 -), carpenters of various generations were recruited to take part in the project, so as to transfer timber processing and bridge building technologies to younger generations.

To foster future carpenters, the workers who were engaged in the most recent project are currently organizing various training programmes, including seminars. The City of Iwakuni believes that such programmes are essential for handing down the traditional technology to future generations. With this view, the City is considering establishing an appropriate system to support such activities.

Timber Production System

During the Edo Period, the Bridge was built using locally available timber. In recent projects, however, including the one during the Heisei Period (1989 -), rebuilding projects had to rely primarily on a supply of timber from other prefectures. There are various, rigorous requirements for the wood materials to be used in building the Bridge, since they must withstand extremely severe climatic and weather conditions. Only timbers of large diameter can meet these requirements. In the local mountains and forests owned by the City of Iwakuni, however, no such trees were found.

The City of Iwakuni therefore has designated a special forest to grow wood for use in future Kintaikyo Bridge rebuilding projects. Since the trees planted in the forest are still young - about 80 years old -, they cannot be used now. They should be preserved, however, for use in future. To pass down the traditional bridge building technology to future generations, some suggested that the Bridge be rebuilt every twenty years.

In line with this proposal, the authority has designated a new forest, where seedlings will be planted in March 2008 to prepare timbers for a project 200 years hence. Such a timber production system, however, is not new. There is a document indicating that during the Edo Period, the Iwakuni Domain planted trees for use in the Bridge. More recently, 2,000 zelkova trees were planted in 1991.

The tree-planting campaign for the Bridge has many purposes: preserving and passing down local traditional culture and technology, nourishing water sources, helping alleviate global warming, preserving biodiversity and developing a local culture of preserving the historic environment through Bridge rebuilding and tree planting on a regular basis.

Rebuilding Expenses

Since the Iwakuni Domain needed considerable funds to build and rebuild the Bridge in 1673 and 1674, respectively, the Domain ordered its vassals and merchants to share the expense. The share for vassals was determined in accordance with their positions; the share of merchants was based on their family businesses. This system continued until 1871. Over the subsequent 95 years, Bridge rebuilding and maintenance was funded by donations and taxes.

On April 1, 1966 Iwakuni City began collecting tolls from pedestrians who cross the Bridge. The project to rebuild the Bridge during the Heisei Period cost 2.6 billion yen. Of the amount, 2.2 billion yen was funded by the tolls, the remaining 0.4 billion yen by the national and prefectural governments. To maintain the Kintaikyo Bridge, the local authority will continue collecting tolls.

Prefabricated Bridge Elements and Systems in Japan and Europe

The aging highway bridge infrastructure in the United States is being subjected to increasing traffic volumes and must be continuously renewed while accommodating traffic flow. The travelling public is demanding that this rehabilitation and replacement be done more quickly to reduce congestion and improve safety. Conventional bridge reconstruction is typically on the critical path because of the sequential, labour-intensive processes of completing the foundation, the substructure, the superstructure components (girders and decks), the railings, and other accessories.

New bridge systems are needed that will allow components to be fabricated off-site and moved into place for quick assembly while maintaining traffic flow. Depending on the specific site conditions, the use of prefabricated bridge systems can minimize traffic disruption, improve work-zone safety, minimize disruption to the environment, improve constructibility, increase quality, and lower life-cycle costs. This technology is applicable and needed for both existing and new bridge construction. The focus of this initiative is on conventional routine bridges that make up the majority of the bridges in the United States.

To obtain information about technologies being used in other industrialized countries, a scanning tour of five countries was made in April 2004. The overall objectives of the scanning tour were to identify international uses of prefabricated bridge elements and systems and to identify decision processes, design methodologies, construction techniques, costs, and maintenance and inspection issues associated with use of the technology. The scanning team was, therefore, interested in all aspects of design, construction, and maintenance of bridge systems composed of multiple elements that are fabricated and assembled off-site. The elements consisted of foundations, piers or columns, abutments, pier caps, beams or girders, and decks. Bridges with span lengths in the range of 6 to 40 metres (20 to 140 ft) were the major focus, although longer spans were of interest if a large amount of innovative prefabrication was used.

The focus areas of the study were, therefore, prefabricated bridge systems that

1. minimize traffic disruption,
2. improve work zone safety,
3. minimize environmental impact,
4. improve constructibility,
5. increase quality, and
6. lower life-cycle costs.

The scanning tour was sponsored by the Federal Highway Administration (FHWA) and the American Association of State Highway and Transportation Officials (AASHTO) and organized by American Trade Initiatives, Inc. The eleven member team included three representatives from FHWA, four representatives from State DOTs, one representative from county engineers, one university representative, and two representatives from industry. The panel visited Japan, the Netherlands, Belgium, Germany, and France and held meetings and site visits with representatives of government agencies and private sector organizations. The countries were selected because of their known use of prefabricated systems. Visiting Japan was particularly important because of their seismic design requirements.

Summary of Initial Findings and Recommendations

At completion of the scanning tour, the team had identified 35 bridge technologies that, in one or more aspects, were different from current practices in the United States. Not all of these related to the primary objectives of the scanning tour. Using the six focus areas as selection criteria, the team identified nine overall technologies that are initially recommended for possible implementation in the United States.

Superstructure Systems

The typical sequence of erecting bridge superstructures in the United States is to erect the concrete or steel beams; place temporary formwork or stay-in-place steel or concrete panels, place deck reinforcement, cast deck concrete; and remove formwork if necessary. Elimination of the need to place and remove formwork for the deck can accelerate construction and improve safety. Two systems to accomplish this were identified during the tour.

Partial Depth Concrete Decks Prefabricated on Steel or Concrete Beams

One system in Germany involved the casting of partial depth concrete decks on steel or concrete beams prior to erection of the beams. After the beams are erected, the edges of each deck unit abut the adjacent member and there is no need to place additional formwork for the cast-in-place concrete. This process speeds construction and reduces the potential danger of equipment falling onto the roadway below since a safe working surface is available immediately after beam erection.

Poutre Dalle System

Another method to eliminate formwork and provide a working surface is provided by a system known as the Poutre Dalle system. In this system, inverted T-beams are placed adjacent to each other and then made composite with cast-in-place concrete placed between the webs of the tees and over the tops of the stems to form a solid member.

Deck Systems

Four innovative ideas for bridge deck systems were identified and are recommended for implementation in the United States.

Full Depth Prefabricated Concrete Decks: The use of full depth prefabricated concrete decks reduces construction time by eliminating the need to erect formwork and provide cast-in-place concrete. The deck panels are connected to steel beams through the use of studs located in pockets in the concrete deck slab. The use of full depth prefabricated concrete decks provides a means to accelerate bridge construction using a factory produced product.

Hybrid Steel-Concrete Deck Systems: The Japanese have developed hybrid steel-concrete slab systems for bridge decks. This steel component of the system consists of transverse beams that span between the longitudinal girders and cantilever beyond the edge girders.

The bottom flanges of the transverse beams support steel formwork for the bottom of the slab while the top flanges support the longitudinal deck reinforcement. When filled with cast-in-place concrete, the system acts as a composite deck system. The system allows rapid placement of a lightweight deck formwork system complete with reinforcement without the use of heavy lifting equipment.

Deck Joint Closure Details: Prefabricated deck systems require that longitudinal and transverse joints be provided to make the deck continuous for live load distribution. This is accomplished by using special loop bar reinforcement details in the joints. Various joint details observed during the tour should be developed for use in the United States to facilitate the use of prefabricated full depth deck systems.

Multiple Level Corrosion Protection Systems: In Japan, Germany, and France, concrete bridge decks are covered with a multiple level corrosion protection system to prevent the ingress of water and deicing chemicals. The systems generally involve providing adequate concrete cover to the reinforcement, a concrete sealer, waterproof membrane, and two layers of asphalt. This type of corrosion protection system may be beneficial with prefabricated systems as a means of protecting the joint regions from potential corrosion damage, thereby ensuring a longer service life.

Substructure Systems

One substructure system is recommended for implementation in the United States.

The SPER System: The SPER system is a method of rapid construction of piers using precast concrete panels as both structural elements and formwork for cast-in-place concrete. Tall hollow piers use panels for inner and outer formwork, while shorter solid piers use panels for the outer formwork only. Testing has shown that the system provides similar seismic resistance as a conventional cast-in-place system. The system has the advantage of more rapid construction with improved constructibility and quality.

Movement Systems

During the tour, many different methods that can be used to remove partial or complete existing bridges and move bridge components or complete bridges into place were observed. These methods allow a new bridge to be built at one location near or adjacent to the existing structure and then moved to its final location in a short period of time. Construction can, therefore, take place in an environment where construction operations are completely separated from the travelling public.

These methods reduce traffic disruption times from months to hours or days, restore the use of existing highways in significantly less time, improve work zone safety, minimize environmental impact, and improve constructibility. This concept of building bridges off line and then moving them into place needs to be developed for use in the United States. Various systems for moving bridges and their components were identified during the tour.

Self Propelled Modular Trailers: In Europe, it was observed that large bridge components or even complete bridges weighing several thousand metric tons have been built at one location and then lifted and transported to their final location using a series of vehicles known as self propelled modular trailers (SPMTs). These multi-axle computer-controlled vehicles have the capability of moving in any horizontal direction while maintaining their payload geometry undeformed or distorted, horizontally, and keeping equal axle loads.

Other Bridge Installation Systems: In addition to using conventional land and/or barge mounted cranes to erect large structures, special load frames have been developed to erect prefabricated components. One example is the modular jacking system used to erect prefabricated pylons on the Millau cable stayed bridge in France. The other methods of moving bridge components included the following:

1. horizontally skidding or sliding bridges into place,
2. incremental launching longitudinally over valleys or above existing highways,
3. floating bridges into place using barges or by building a temporary dry dock,
4. building bridges alongside an existing roadway and rotating them into place, and
5. vertically lifting bridges.

These systems can be used to minimize the time that an existing bridge is out of service while it is replaced, many within 3 to 48 hours.

New Construction Method for Arch Bridges

A Japanese builder is using a new construction method called concrete lapping with preerected composite arch (CLCA). CLCA eliminates the shortcomings of the traditional arch construction methods and meets stringent cost and scheduling requirements. P. S. Concrete Ltd. of Tokyo has used CLCA to construct two bridges in Japan - the Asahi Bridge and the Joshi Bridge. Both bridges took two 8-month construction seasons to build.

CLCA uses rectangular cross-section tubular steel arches filled with concrete to act as the false work to form the reinforced concrete bridge arch. The first step in the process is to construct reinforced concrete abutments on each side of the river. Next, an assembly consisting of several tubular steel half arches is built standing upright on its end on opposite sides of the river. Each half is swung down around hinges at its springings to join at the arch crown, then the joined steel arch tubes are pumped full of concrete. Reinforced steel is placed around the composite structures initally using conventional techniques, and finally using movable platforms that work their way up the composite arches from each side of the river. The bridge is finished using conventional methods.

Bridge Maintenance and Earthquake Protection

Figure: *Minato Bridge*

Many of the large bridges constructed during the heyday of rapid economic growth in Japan are now starting to age, making infrastructure maintenance an issue of increasing concern.

Major Technologies

Preventative Maintenance

At a time when much infrastructure is aging, it is increasingly important to develop preventative maintenance regimes involving early detection and repair of faults and damage in order to minimize overall life cycle costs and extend the usable life of bridges. Preventative maintenance is based on the concept of asset management predicated on efficient and effective maintenance programmes.

The Hitachi Zosen Bridge Management System brings together the combined expertise of the entire Hitachi Zosen group in a wide range of fields under a systematic framework encompassing inspection, evaluation and damage forecasting systems, repair and reinforcement techniques, renovation and upgrading solutions, and total bridge replacement.

Inspection, Monitoring and Emergency Systems

Hitachi Zosen is developing a bridge maintenance database based on the three-dimensional product models currently used at the factory fabrication stage. Under this approach, laser scanners or equivalent are used to generate 3-D measurement data that is combined with product models to generate a comprehensive database to assist in the development of bridge maintenance solutions.

Hitachi Zosen is also working on disaster prevention and emergency systems that combine GPS technology (which played a vital role in providing measurements of tsunamis during the Great East Japan Earthquake) with the Geographic Information System (GIS) and can be used to generate comprehensive route information, monitor the structural integrity of bridges, and provide communication services at times of emergency.

Seismic Analysis

Seismic reinforcement of very large bridges is shifting towards damage control designs using devices that are designed to reduce seismic forces more efficiently by absorbing earthquake energy. This approach necessitates highly advanced dynamic analysis techniques.

The Hitachi Zosen group boasts extensive experience and expertise in dynamic response analysis and FEM analysis under special conditions on a wide range of different bridge types, from very large bridges to ordinary bridges. We have applied our knowledge to major reinforcement works on very large bridges such as the Minato and Tempozan Bridges in Osaka.

Renovation and Upgrading

Many aging bridges suffer not only from deterioration and damage over time but also from higher traffic levels, heavier truck loads, environmental impacts such as ground subsidence and the need for improved seismic reinforcement. The Hitachi Zosen group refurbishes aging bridges using a range of techniques such as seismic reinforcement and replacement of floor systems that can extend the usable life of bridges and allow them to serve as important public assets for many more years and carry on into the next generation.

Hitachi Zosen was the first in Japan to apply the cable tension readjustment technique (on the Kamome Bridge) and has upgraded many bridges by replacing the floor system.

7

Seismic Isolation and Supplemental Energy Dissipation

Seismic isolation and energy dissipation systems are viable design strategies that have already been used for seismic rehabilitation of a number of buildings. Other special seismic protective systems—including active control, hybrid combinations of active and passive energy devices, and tuned mass and liquid dampers—may also provide practical solutions in the near future.

These systems are similar in that they enhance performance during an earthquake by modifying the building's response characteristics. Seismic isolation and energy dissipation systems will not be appropriate design strategies for most buildings, particularly buildings that have only Limited

Rehabilitation Objectives. In general, these systems will be most applicable to the rehabilitation of buildings whose owners desire superior earthquake performance and can afford the special costs associated with the design, fabrication, and installation of seismic isolators and/or energy dissipation devices. These costs are typically offset by the reduced need for stiffening and strengthening measures that would otherwise be required to meet Rehabilitation Objectives.

Seismic isolation and energy dissipation systems are relatively new and sophisticated concepts that require more extensive design and detailed analysis than do most conventional rehabilitation schemes. Similarly, design (peer) review is required for all rehabilitation schemes that use either seismic isolation or energy dissipation systems. Seismic isolation and energy dissipation systems include a wide variety of concepts and devices. In most cases, these systems and devices will be implemented with some additional conventional strengthening of the structure; in all cases they will require evaluation of existing building elements.

Seismic isolation is increasingly being considered for historic buildings that are free-standing and have a basement or bottom space of no particular historic significance. In selecting such a solution, special consideration should be given to the possibility that historic or archaeological resources may be present at the site. If historic or archaeological resources are present at the site, the guidance of the State Historic Preservation Officer should be obtained in a timely manner. Isolation is also often considered for essential facilities, to protect valuable contents, and on buildings with a complete but insufficiently strong lateral-force-resisting system.

Conceptually, isolation reduces response of the superstructure by "decoupling" the building from the ground. Typical isolation systems reduce forces transmitted to the superstructure by lengthening the period of the building and adding some amount of damping. Added damping is an inherent property of most isolators, but may also be provided by supplemental energy dissipation devices installed across the isolation interface. Under favourable conditions, the isolation system reduces drift in the superstructure by a factor of at least two—and sometimes by as much as factor of five—from that which would occur if the building were not isolated.

Accelerations are also reduced in the structure, although the amount of reduction depends on the force-deflection characteristics of the isolators and may not be as significant as the reduction of drift. Reduction of drift in the superstructure protects structural components and elements, as well as nonstructural components sensitive to drift-induced damage.

Reduction of acceleration protects nonstructural components that are sensitive to acceleration-induced damage. Passive energy dissipation devices add damping (and sometimes stiffness) to the building. A wide variety of passive energy dissipation devices are available, including fluid viscous dampers, viscoelastic materials, and hysteretic devices. Ideally, energy dissipation devices dampen earthquake excitation of the structure that would otherwise cause higher levels of response and damage to components of the building. Under favourable conditions, energy dissipation devices reduce drift of the structure by a factor of about two to three (if no stiffness is added) and by larger factors if the devices also add stiffness to the structure.

Energy dissipation devices will also reduce force in the structure—provided the structure is responding elastically—but would not be expected to reduce force in structures that are responding beyond yield.

Active control systems sense and resist building motion, either by applying external force or by modifying structural properties of active elements (e.g., so-called "smart" braces). Tuned mass or liquid dampers modify properties and add damping to key building modes of vibration. Special seismic systems, such as isolation or energy dissipation systems, should be considered early in the design process and be based on the Rehabilitation Objectives established for the building.

Whether a special seismic system is found to be the "correct" design strategy for building rehabilitation. will depend primarily on the performance required at the specified level of earthquake demand. In general, special seismic systems will be found to be more attractive as a rehabilitation strategy for buildings that have more stringent Rehabilitation Objectives (i.e., higher levels of performance and more severe levels of earthquake demand).

Seismic Isolation Systems

To many structural engineers, the conventional approach to protect buildings from the destructive forces of earthquakes is to increase the strength of the buildings so that they do not collapse during such events. This approach is not entirely effective in terms of protection afforded to the contents and occupants as a result of the application of the forces transmitted in to the building.

Since the motion of earthquakes is vibrational in nature, the principle of vibration isolation can be utilized to protect a building is decoupled from the horizontal components of the earthquake ground motion by mounting rubber bearings between the building and its foundation. Such a system not only provides protection to the building but to its contents and occupants as well.

Vibration Isolation

Vibration isolation is the process of isolating an object, such as a piece of equipment, from the source of vibrations.

Passive Isolation

"Passive vibration isolation" refers to vibration isolation or mitigation of vibrations by passive techniques such as rubber pads or mechanical springs, as opposed to "active vibration isolation" or "electronic force cancellation" employing electric power, sensors, actuators, and control systems.

Passive vibration isolation is a vast subject, since there are many types of passive vibration isolators used for many different applications. A few of these applications are for industrial equipment such as pumps, motors, HVAC systems, or washing machines; isolation of civil engineering structures from earthquakes (base isolation), sensitive laboratory equipment, valuable statuary, and high-end audio.

A basic understanding of how passive isolation works, the more common types of passive isolators, and the main factors that influence the selection of passive isolators:

Common Passive Isolation Systems

Pneumatic or air Isolators: These are bladders or canisters of compressed air. A source of compressed air is required to maintain them. Air springs are rubber bladders which provide damping at the same time as isolation and are used in large trucks. Some pneumatic isolators can attain low resonant frequencies and are used for isolating large industrial equipment. Air tables consist of a working surface or optical surface mounted on air legs. These tables provide enough isolation for laboratory instrument under some conditions. Air systems may leak under vacuum conditions. The air container can interfere with isolation of low-amplitude vibration.

Mechanical Springs and Spring-dampers: These are heavy-duty isolators used for building systems and industry. Sometimes they serve as mounts for a concrete block, which provides further isolation.

Pads or Sheets of Flexible Materials Such as Elastomers, Rubber, Cork, etc.: Cork and/or elastomer pads are often used under heavy machinery or even under audio systems.

Molded and Bonded Rubber and Elastomeric Isolators and Mounts: These are often used as machinery mounts or in vehicles. They absorb shock and attenuate some vibration.

Negative-stiffness Isolators: Negative-stiffness isolators are less common than other types and have generally been developed for high-level research applications such as gravity wave detection. Lee,

Goverdovskiy, and Temnikov (2007) proposed a negative-stiffness system for isolating vehicle seats.

The focus on negative stiffness isolators has been on developing systems with very low resonant frequencies (below 1 Hz), so that low frequencies can be adequately isolated, which is critical for sensitive instrumentation.

In addition, all higher frequencies are also isolated. Negative stiffness systems can be made with low stiction, so that they are effective in isolating low-amplitude vibrations.

Negative-stiffness mechanisms are purely mechanical and typically involve the configuration and loading of components such as beams or inverted pendulums. Greater loading of the negative-stiffness mechanism, within the range of its operability, decreases the natural frequency.

Wire Rope Isolators: These isolators are durable and can withstand extreme environments. They are often used in military applications.

Bungee Cord Isolators and Tennis Balls: Bungee cords can be used as a cheap isolation system which may be effective enough for some applications. The item to be isolated is suspended from the bungee cords. This is difficult to implement without a danger of the isolated item falling. Tennis balls cut in half have been used under washing machines and other items with some success.

Base Isolators for Seismic Isolation of Buildings, Bridges, etc.: Base isolators made of layers of neoprene and steel with a low horizontal stiffness are used to lower the natural frequency of the building. Some other base isolators are designed to slide, preventing the transfer of energy from the ground to the building.

Tuned Mass Dampers: Tuned mass dampers reduce the effects of harmonic vibration in buildings or other structures. A relatively small mass is attached in such a way that it can dampen out a very narrow band of vibration of the structure

How Passive Isolation Works

A passive isolation system in general contains mass, spring, and damping elements and moves as a harmonic oscillator. The mass and spring stiffness dictate a natural frequency of the system. Damping causes energy dissipation and has a secondary effect on natural frequency.

Every object on a flexible support has a fundamental natural frequency. When vibration is applied, energy is transferred most efficiently at the natural frequency, somewhat efficiently below the natural frequency, and with increasing inefficiency (decreasing efficiency) above the natural frequency. This can be seen in the transmissibility curve, which is a plot of transmissibility vs. frequency.

Here is an example of a transmissibility curve. Transmissibility is the ratio of vibration of the isolated surface to that of the source. Vibrations are never completely eliminated, but they can be greatly reduced. The curve below shows the typical performance of a passive, negative-stiffness isolation system with a natural frequency of 0.5 Hz. The general shape of the curve is typical for passive systems. Below the natural frequency, transmissibility hovers near 1.

A value of 1 means that vibration is going through the system without being amplified or reduced. At the resonant frequency, energy is transmitted efficiently, and the incoming vibration is amplified. Damping in the system limits the level of amplification. Above the resonant frequency, little energy can be transmitted, and the curve rolls off to a low value. A passive isolator can be seen as a mechanical low-pass filter for vibrations.

In general, for any given frequency above the natural frequency, an isolator with a lower natural frequency will show greater isolation than one with a higher natural frequency. The best isolation system for a given situation depends on the frequency, direction, and magnitude of vibrations present and the desired level of attenuation of those frequencies.

All mechanical systems in the real world contain some amount of damping. Damping dissipates energy in the system, which reduces the vibration level which is transmitted at the natural frequency. The fluid in automotive shock absorbers is a kind of damper, as is the inherent damping in elastomeric (rubber) engine mounts.

Damping is used in passive isolators to reduce the amount of amplification at the natural frequency. However, increasing damping tends to reduce isolation at the higher frequencies. As damping is increased, transmissibility roll-off decreases. This can be seen in the chart below.

Passive isolation operates in both directions, isolating the payload from vibrations originating in the support, and also isolating the support from vibrations originating in the payload. Large machines such as washers, pumps, and generators, which would cause vibrations in the building or room, are often isolated from the floor. However, there are a multitude of sources of vibration in buildings, and it is often not possible to isolate each source. In many cases, it is most efficient to isolate each sensitive instrument from the floor. Sometimes it is necessary to implement both approaches.

Factors Influencing the Selection of Passive Vibration Isolators

1. Characteristics of item to be isolated
   - o Size: The dimensions of the item to be isolated help determine the type of isolation which is available and appropriate. Small objects may use only one isolator, while larger items might use a multiple-isolator system.
   - o Weight: The weight of the object to be isolated is an important factor in choosing the correct passive isolation product. Individual passive isolators are designed to be used with a specific range of loading.
   - o Movement: Machines or instruments with moving parts may affect isolation systems. It is important to know the mass, speed, and distance travelled of the moving parts.
2. Operating Environment
   - o Industrial: This generally entails strong vibrations over a wide band of frequencies and some amount of dust.
   - o Laboratory: Labs are sometimes troubled by specific building vibrations from adjacent machinery, foot traffic, or HVAC airflow.
   - o Indoor or outdoor: Isolators are generally designed for one environment or the other.
   - o Corrosive/non-corrosive: Some indoor environments may present a corrosive danger to isolator components due to the presence of corrosive chemicals. Outdoors, water and salt environments need to be considered.
   - o Clean room: Some isolators can be made appropriate for clean room.
   - o Temperature: In general, isolators are designed to be used in the range of temperatures normal for human environments. If a larger range of temperatures is required, the isolator design may need to be modified.

- o Vacuum: Some isolators can be used in a vacuum environment. Air isolators may have leakage problems. Vacuum requirements typically include some level of clean room requirement and may also have a large temperature range.
- o Magnetism: Some experimentation which requires vibration isolation also requires a low-magnetism environment. Some isolators can be designed with low-magnetism components.
- o Acoustic noise: Some instruments are sensitive to acoustic vibration. In addition, some isolation systems can be excited by acoustic noise. It may be necessary to use an acoustic shield. Air compressors can create problematic acoustic noise, heat, and airflow.
- o Static or dynamic loads: This distinction is quite important as isolators are designed for a certain type and level of loading.

Static loading is basically the weight of the isolated object with low-amplitude vibration input. This is the environment of apparently stationary objects such as buildings (under normal conditions) or laboratory instruments.

Dynamic loading involves accelerations and larger amplitude shock and vibration. This environment is present in vehicles, heavy machinery, and structures with significant movement.

3. Cost:
   - o Cost of providing isolation: Costs include the isolation system itself, whether it is a standard or custom product; a compressed air source if required; shipping from manufacturer to destination; installation; maintenance; and an initial vibration site survey to determine the need for isolation.
   - o Relative costs of different isolation systems: Inexpensive shock mounts may need to be replaced due to dynamic loading cycles. A higher level of isolation which is effective at lower vibration frequencies and magnitudes generally costs more. Prices can range from a few dollars for bungee cords to millions of dollars for some space applications.
4. Adjustment: Some isolation systems require manual adjustment to compensate for changes in weight load, weight distribution,

temperature, and air pressure, whereas other systems are designed to automatically compensate for some or all of these factors.

5. Maintenance: Some isolation systems are quite durable and require little or no maintenance. Others may require periodic replacement due to mechanical fatigue of parts or aging of materials.
6. Size Constraints: The isolation system may have to fit in a restricted space in a laboratory or vacuum chamber, or within a machine housing.
7. Nature of vibrations to be isolated or mitigated
   - o Frequencies: If possible, it is important to know the frequencies of ambient vibrations. This can be determined with a site survey or accelerometer data processed through FFT analysis.
   - o Amplitudes: The amplitudes of the vibration frequencies present can be compared with required levels to determine whether isolation is needed. In addition, isolators are designed for ranges of vibration amplitudes. Some isolators are not effective for very small amplitudes.
   - o Direction: Knowing whether vibrations are horizontal or vertical can help to target isolation where it is needed and save money.
8. Vibration specifications of item to be isolated: Many instruments or machines have manufacturer-specified levels of vibration for the operating environment. The manufacturer may not guarantee the proper operation of the instrument if vibration exceeds the spec.

Comparison of Passive Isolators

Negative-stiffness Vibration Isolator

Negative-Stiffness-Mechanism (NSM) vibration isolation systems offer a unique passive approach for achieving low vibration environments and isolation against sub-Hertz vibrations. "Snap-through" or "over-centre" NSM devices are used to reduce the stiffness of elastic suspensions and create compact six-degree-of-freedom systems with low natural frequencies. Practical systems with vertical and horizontal natural frequencies as low as 0.2 to 0.5 Hz are possible.

Electro-mechanical auto-adjust mechanisms compensate for varying weight loads and provide automatic leveling in multiple-isolator systems, similar to the function of levelling valves in pneumatic systems. All-metal systems can be configured which are compatible with high vacuums and other adverse environments such as high temperatures.

These isolation systems enable vibration-sensitive instruments such as scanning probe microscopes, micro-hardness testers and scanning electron microscopes to operate in severe vibration environments sometimes encountered, for example, on upper floors of buildings and in clean rooms. Such operation would not be practical with pneumatic isolation systems. Similarly, they enable vibration-sensitive instruments to produce better images and data than those achievable with pneumatic isolators.

The theory of operation of NSM vibration isolation systems is summarized, some typical systems and applications are described, and data on measured performance is presented. The theory of NSM isolation systems is explained in References 1 and 2. It is summarized briefly for convenience.

Vertical-motion Isolation

A vertical-motion uses a conventional spring connected to an NSM consisting of two bars hinged at the centre, supported at their outer ends on pivots, and loaded in compression by forces P. The spring is compressed by weight W to the operating position of the isolator. The stiffness of the isolator is K=KS-KN where KS is the spring stiffness and KN is the magnitude of a negative stiffness which is a function of the length of the bars and the load P. The isolator stiffness can be made to approach zero while the spring supports the weight W.

Horizontal-motion Isolation

A horizontal-motion isolator consisting of two beam-columns. Each beam-column behaves like two fixed-free beam columns loaded axially by a weight load W. Without the weight load the beam-columns have horizontal stiffness KS With the weight load the lateral bending stiffness is reduced by the "beam-column" effect. This behaviour is equivalent to a horizontal spring combined with an NSM so that the horizontal stiffness is K=KS-KN, and KN is the magnitude of the beam-column effect.

Horizontal stiffness can be made to approach zero by loading the beam-columns to approach their critical buckling load.

Six-Degree-of-Freedom (six-DOF) isolation

A six-DOF NSM isolator typically uses three isolators stacked in series: a tilt-motion isolator on top of a horizontal-motion isolator on top of a vertical-motion isolator. A schematic of a vibration isolation system consisting of a weighted platform supported by a single six-DOF isolator incorporating the isolators. Flexures are used in place of the hinged bars. A tilt flexure serves as the tilt-motion isolator. A vertical-stiffness adjustment screw is used to adjust the compression force on the negative-stiffness flexures thereby changing the vertical stiffness. A vertical load adjustment screw is used to adjust for varying weight loads by raising or lowering the base of the support spring to keep the flexures in their straight, unbent operating positions.

Vibration Isolation of Supporting Joint

The equipment and gears have joint with surrounding objects (the supporting joint - with the support; the unsupporting joint - the pipe duct or cable). Vibration isolation of supporting joint is realized in the device named vibration-isolator (absorber). On an illustration presented dependence of difference is levels of vibrations which are measured before installation of the functioning gear on vibration-isolator and after installation in a wide range of frequencies.

The Vibration Isolator

A device that reflects and absorbs waves of oscillatory energy, extending from the working gear or an electrical equipment, with the aid of effect of a vibration insulation. Vibration-isolator is established between a body transferring fluctuations and a body which defend (for example, between the gear and tho foundation). In a structure is used the rubber envelope, which is reinforcod by a spring. Rubber and a spring are strongly connected during transformation of crude rubber into rubber envelope by a method of vulcanization. Under action of weight loading of the gear the rubber envelope is of deformation, and a spring are compressed or stretch.

Thus, in springs cross section, occurs the twig twist with a material of rubber envelope, causing doformation of shift in rubber envelope. It is known, that the vibration insulation basically cannot be carried out without presence of vibration absorption. The size of deformation of shift in elastic material of vibration-isolator it basis for definition of size of absorption of fluctuations. At action of vibration or shock loadings of deformation increase. Being thus cyclic, it considerably strengthens efficiency of the given device. In the upper part of a design the sleeve, and in the lower part a flange by means of which the vibration-isolator fastens to the gear and the foundation.

Vibration Isolation of Unsupporting Joint

Vibration isolation of unsupporting joint is realized in the device named branch pipe a of isolating vibration.

Branch Pipe a of Isolating Vibration

Branch pipe a of isolating vibration is a part of a tube with elastic walls for reflection and absorption of waves of the oscillatory energy extending from the working pump over wall of the pipe duct. Is established between the pump and the pipe duct. In a structure is used the rubber envelope, which is reinforced by a spring. Properties of an envelope are similar envelope to a isolator vibration. Has the device reducing axial effort from action of internal pressure up to zero.

Subframe Isolation

Another technique used to increase isolation is to use an isolated subframe. This splits the system with an additional mass/spring/ damper system. This doubles the high frequency attenuation rolloff, at the cost of introducing additional low frequency modes which may cause the low frequency behaviour to deteriorate. This is commonly used in the rear suspensions of cars with Independent Rear Suspension (IRS), and in the front subframes of some cars. The graph shows the force into the body for a subframe that is rigidly bolted to the body compared with the red curve that shows a compliantly mounted subframe. Above 42 Hz the compliantly mounted subframe is superior, but below that frequency the bolted in subframe is better.

Bushing

A bushing or rubber bushing is a type of vibration isolator. It provides an interface between two parts, damping the energy transmitted through the bushing. A common application is in vehicle suspension systems, where a bushing made of rubber (or, more often, synthetic rubber or polyurethane) separates the faces of two metal objects while allowing a certain amount of movement. This movement allows the suspension parts to move freely, for example, when travelling over a large bump, while minimizing transmission of noise and small vibrations through to the chassis of the vehicle. A rubber bushing may also be described as a flexible mounting or antivibration mounting.

These bushings often take the form of an annular cylinder of flexible material inside a metallic casing or outer tube. They might also feature an internal *crush tube* which protects the bushing from being crushed by the fixings which hold it onto a threaded spigot. Many different types of bushing designs exist. An important difference compared with plain bearings is that the relative motion between the two connected parts is accommodated by strain in the rubber, rather than by shear or friction at the interface. Some rubber bushings, such as the D block for a sway bar, do allow sliding at the interface between one part and the rubber.

Advantages and Disadvantages

The main advantage of a bushing, as compared to a solid connection, is less noise and vibration are transmitted. Another advantage is that they require little to no lubrication.

Disadvantages include:

- Rubber bushings can deteriorate quickly in the presence of oils (e.g., motor oil, mineral oil) and extreme heat and cold.
- The flexibility of rubber also introduces an element of play in the suspension system. This may result in camber, caster, or toe changes in the wheels of the vehicle during high-load conditions (cornering and braking), adversely affecting the vehicle's handling. For this reason, a popular aftermarket performance upgrade is the replacement of rubber suspension bushings with bushings made of more rigid materials, such as polyurethane. Polyurethane bushings are also available for many vehicles with approximately the same characteristics as the manufacturers original bushings, but with greatly increased durability. This is useful on vehicles that have a reputation for wearing out standard rubber bushings, but for which harder bushings with increased harshness of ride are not wanted.

Applications

- In vehicles:
  - o Sway bar links and mountings
  - o Shock absorber mountings
  - o Double wishbone suspension assemblies

- o Gear stick for vehicles with a manual transmission
- o Most high-speed inline internal combustion engines are prone to torsional vibration of their crankshafts; the straight six and straight eight engines being particularly prone to this problem due to their long crankshaft length. Although straight eight engines faded from the marketplace in the 1950s, many straight six engines have and still do feature crankshaft vibration damping utilizing rubber bushes. The 3,442 cc Jaguar XK 6-cylinder engine of 1948 and most subsequent versions of the ubiquitous Jaguar XK engine used a proprietary Metalastik vibration damper to protect their crankshafts from potentially damaging torsional vibrations. To quote William Heynes, "The Metalastik damper consists of a steel plate to which is bonded, through a thick rubber disk, a malleable iron floating weight. Variations of the weight, rubber volume and mix, give these dampers a very wide field over which they can operate."

- In skateboards, bushings enable the tilting of the trucks, thus allowing the board to turn.
- In fastening, bushings are also used to transfer loads from a fastening to a much larger area in the underlying structure, the object being to reduce the strain on individual fibres within the underlying structure.

Project Inception

Engineering projects are, in many countries, specifically defined by legislation, which requires that such projects should be carried out by registered engineers and/or registered engineering companies. That is, companies with license to carry out such works as design and construction of buildings, power plants, industrial facilities, installation and erection of electrical grid networks, transportation infrastructure and the like.

The scope of the project is specified in a contract between the owner and the engineering and construction parties. As a rule, an engineering project is broken down into design and construction phases. The outputs of the design process are drawings, calculations, and all other design documentation necessary to carry out the next phase. The next phase would normally be sending the project plans to a developer who will then help construct the plans (construction phase).

Project Management

In project management a project consists of a temporary endeavour undertaken to create a unique product, service or result. Another definition is a management environment that is created for the purpose of delivering one or more business products according to a specified business case.

Project objectives define target status at the end of the project, reaching of which is considered necessary for the achievement of planned benefits. They can be formulated as SMART criteria: Specific, Measurable (or at least evaluable) achievement, Achievable (recently Agreed-to or Acceptable are used regularly as well), Realistic (given the current state of organizational resources) and Time terminated (bounded). The evaluation (measurement) occurs at the project closure. However a continuous guard on the project progress should be kept by monitoring and evaluating. It is also worth noting that SMART is best applied for incremental type innovation projects. For radical type projects it does not apply as well. Goals for such projects tend to be broad, qualitative, stretch/unrealistic and success driven.

The primary challenge of project management is to achieve all of the project goals and objectives while honouring the preconceived constraints. The primary constraints are scope, time, quality and budget. The secondary —and more ambitious— challenge is to optimize the allocation of necessary inputs and integrate them to meet pre-defined objectives.

Until 1900 civil engineering projects were generally managed by creative architects, engineers, and master builders themselves, for example Vitruvius (first century BC), Christopher Wren (1632–1723), Thomas Telford (1757–1834) and Isambard Kingdom Brunel (1806–1859). It was in the 1950s that organizations started to systematically apply project management tools and techniques to complex engineering projects.

As a discipline, project management developed from several fields of application including civil construction, engineering, and heavy defence activity. Two forefathers of project management are Henry Gantt, called the father of planning and control techniques, who is famous for his use of the Gantt chart as a project management tool (alternatively *Harmonogram* first proposed by Karol Adamiecki); and Henri Fayol for his creation of the five management functions that form the foundation of the body of knowledge associated with project and programme management. Both Gantt and Fayol were students of Frederick Winslow Taylor's theories of scientific management. His work is the forerunner to modern project management tools including work breakdown structure (WBS) and resource allocation.

The 1950s marked the beginning of the modern project management era where core engineering fields come together to work as one. Project management became recognized as a distinct discipline arising from the management discipline with engineering model.

In the United States, prior to the 1950s, projects were managed on an ad-hoc basis, using mostly Gantt charts and informal techniques and tools. At that time, two mathematical project-scheduling models were developed. The "Critical Path Method" (CPM) was developed as a joint venture between DuPont Corporation and Remington Rand Corporation for managing plant maintenance projects. And the "Programme Evaluation and Review Technique" or PERT, was developed by Booz Allen Hamilton as part of the United States Navy's (in conjunction with the Lockheed Corporation) Polaris missile submarine programme; These mathematical techniques quickly spread into many private enterprises.

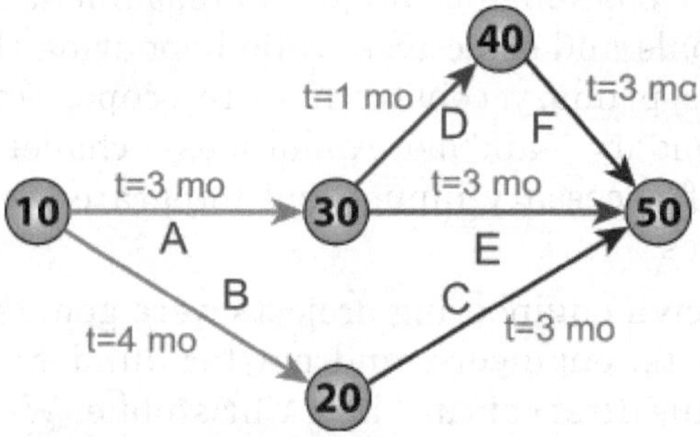

Figure: *PERT network chart for a seven-month project with five milestones*

At the same time, as project-scheduling models were being developed, technology for project cost estimating, cost management, and engineering economics was evolving, with pioneering work by Hans Lang and others.

In 1956, the American Association of Cost Engineers (now AACE International; the Association for the Advancement of Cost Engineering) was formed by early practitioners of project management and the associated specialties of planning and scheduling, cost estimating, and cost/schedule control (project control). AACE continued its pioneering work and in 2006 released the first integrated process for portfolio, programme and project management (Total Cost Management Framework).

The International Project Management Association (IPMA) was founded in Europe in 1967, as a federation of several national project management associations. IPMA maintains its federal structure today and now includes member associations on every continent except Antarctica. IPMA offers a Four Level Certification programme based on the IPMA Competence Baseline (ICB). The ICB covers technical, contextual, and behavioural competencies.

In 1969, the Project Management Institute (PMI) was formed in the USA. PMI publishes *A Guide to the Project Management Body of Knowledge* (PMBOK Guide), which describes project management practices that are common to "most projects, most of the time." PMI also offers multiple certifications.

Approaches

There are a number of approaches to managing project activities including lean, iterative, incremental, and phased approaches. Regardless of the methodology employed, careful consideration must be given to the overall project objectives, timeline, and cost, as well as the roles and responsibilities of all participants and stakeholders.

The Traditional Approach

A traditional phased approach identifies a sequence of steps to be completed. In the "traditional approach", five developmental components of a project can be distinguished (four stages plus control):

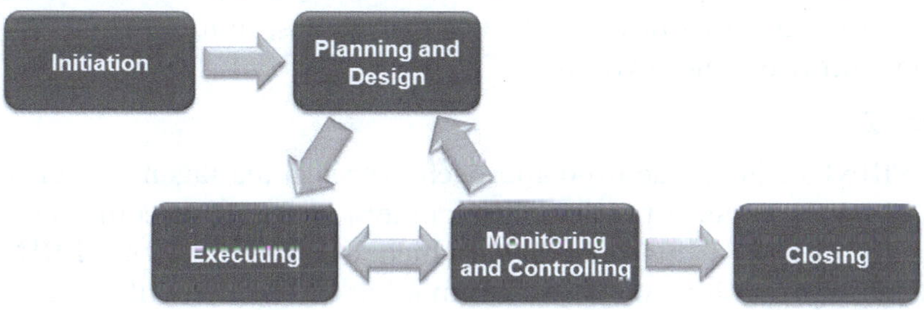

Figure: *Typical development phases of an engineering project*

1. initiation
2. planning and design
3. execution and construction
4. monitoring and controlling systems
5. completion

Not all projects will have every stage, as projects can be terminated before they reach completion. Some projects do not follow a structured planning and/or monitoring process. And some projects will go through steps 2, 3 and 4 multiple times.

Many industries use variations of these project stages. For example, when working on a brick-and-mortar design and construction, projects will typically progress through stages like pre-planning, conceptual design, schematic design, design development, construction drawings (or contract documents), and construction administration. In software development, this approach is often known as the waterfall model, i.e., one series of tasks after another in linear sequence. In software development many organizations have adapted the Rational Unified Process (RUP) to fit this methodology, although RUP does not require or explicitly recommend this practice.

Waterfall development works well for small, well defined projects, but often fails in larger projects of undefined and ambiguous nature. The Cone of Uncertainty explains some of this as the planning made on the initial phase of the project suffers from a high degree of uncertainty. This becomes especially true as software development is often the realization of a new or novel product. In projects where requirements have not been finalized and can change, requirements management is used to develop an accurate and complete definition of the behaviour of software that can serve as the basis for software development. While the terms may differ from industry to industry, the actual stages typically follow common steps to problem solving—"defining the problem, weighing options, choosing a path, implementation and evaluation."

Prince2

PRINCE2 is a structured approach to project management released in 1996 as a generic project management method. It combines the original PROMPT methodology (which evolved into the PRINCE methodology) with IBM's MITP (managing the implementation of the total project) methodology. PRINCE2 provides a method for managing projects within a clearly defined framework.

PRINCE2 focuses on the definition and delivery of products, in particular their quality requirements. As such, it defines a successful project as being output-oriented (not activity- or task-oriented) through creating an agreed set of products that define the scope of the project and provides the basis for planning and control, that is, how then to coordinate people and activities, how to design and supervise product delivery, and what to do if products and therefore the scope of the project has to be adjusted if it does not develop as planned.

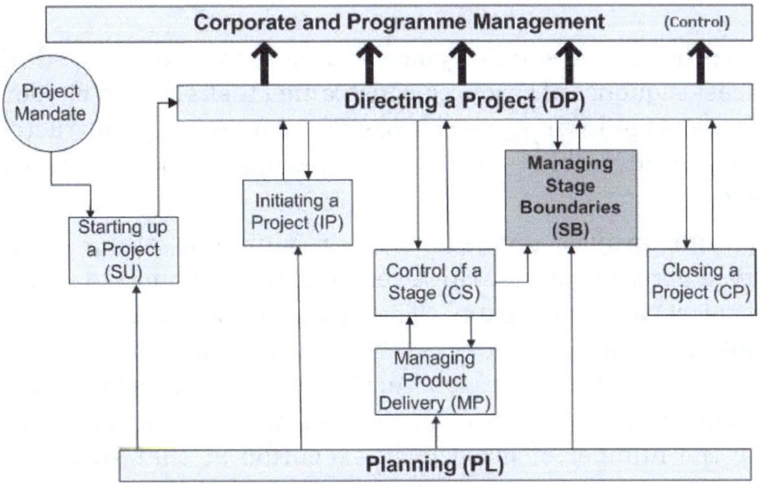

Figure: *The PRINCE2 process model*

In the method, each process is specified with its key inputs and outputs and with specific goals and activities to be carried out to deliver a project's outcomes as defined by its Business Case. This allows for continuous assessment and adjustment when deviation from the Business Case is required.

PRINCE2 provides a common language for all participants in the project. The governance framework of PRINCE2 – its roles and responsibilities – are fully described and require tailoring to suit the complexity of the project and skills of the organisation.

Critical Chain Project Management

Critical chain project management (CCPM) is a method of planning and managing project execution designed to deal with uncertainties inherent in managing projects, while taking into consideration limited availability of resources (physical, human skills, as well as management & support capacity) needed to execute projects.

CCPM is an application of the theory of constraints (TOC) to projects. The goal is to increase the flow of projects in an organization (throughput). Applying the first three of the five focusing steps of TOC, the system constraint for all projects is identified as are the resources. To exploit the constraint, tasks on the critical chain are given priority over all other activities. Finally, projects are planned and managed to ensure that the resources are ready when the critical chain tasks must start, subordinating all other resources to the critical chain.

The project plan should typically undergo resource levelling, and the longest sequence of resource-constrained tasks should be identified as the critical chain. In some cases, such as managing contracted sub-projects, it is advisable to use a simplified approach without resource levelling.

In multi-project environments, resource levelling should be performed across projects. However, it is often enough to identify (or simply select) a single "drum". The drum can be a resource that acts as a constraint across projects, which are staggered based on the availability of that single resource. One can also use a "virtual drum" by selecting a task or group of tasks (typically integration points) and limiting the number of projects in execution at that stage.

Event Chain Methodology

Event chain methodology is another method that complements critical path method and critical chain project management methodologies. Event chain methodology is an uncertainty modelling and schedule network analysis technique that is focused on identifying and managing events and event chains that affect project schedules. Event chain methodology helps to mitigate the negative impact of psychological heuristics and biases, as well as to allow for easy modelling of uncertainties in the project schedules. Event chain methodology is based on the following principles.

- Probabilistic moment of risk: An activity (task) in most real-life processes is not a continuous uniform process. Tasks are affected by external events, which can occur at some point in the middle of the task.
- Event chains: Events can cause other events, which will create event chains. These event chains can significantly affect the course of the project. Quantitative analysis is used to determine a cumulative effect of these event chains on the project schedule.
- Critical events or event chains: The single events or the event chains that have the most potential to affect the projects are the "critical events" or "critical chains of events." They can be determined by the analysis.

- Project tracking with events: Even if a project is partially completed and data about the project duration, cost, and events occurred is available, it is still possible to refine information about future potential events and helps to forecast future project performance.
- Event chain visualization: Events and event chains can be visualized using event chain diagrams on a Gantt chart.

Process-based Management

Also furthering the concept of project control is the incorporation of process-based management. This area has been driven by the use of Maturity models such as the CMMI and ISO/IEC15504 (SPICE – software process improvement and capability estimation).

Agile Project Management

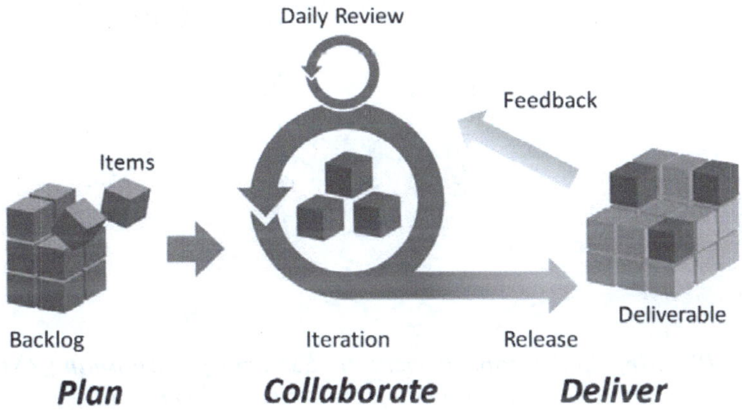

Figure: *The iteration cycle in agile project management*

Agile project management approaches, based on the principles of human interaction management, are founded on a process view of human collaboration. It is "most typically used in software, website, technology, creative and marketing industries."

This contrasts sharply with the traditional approach. In the agile software development or flexible product development approach, the project is seen as a series of relatively small tasks conceived and executed as the situation demands in an adaptive manner, rather than as a completely pre-planned process.

It is the most consistent project management technique since it involves frequent testing of the project under development. It is the only technique in which the client will be actively involved in the project development. But the only disadvantage with this technique is that it should be used only if the client has enough time to be actively involved in the project every now and then.

Lean Project Management

Lean project management uses principles from lean manufacturing to focus on delivering value with less waste and reduced time.

Extreme Project Management

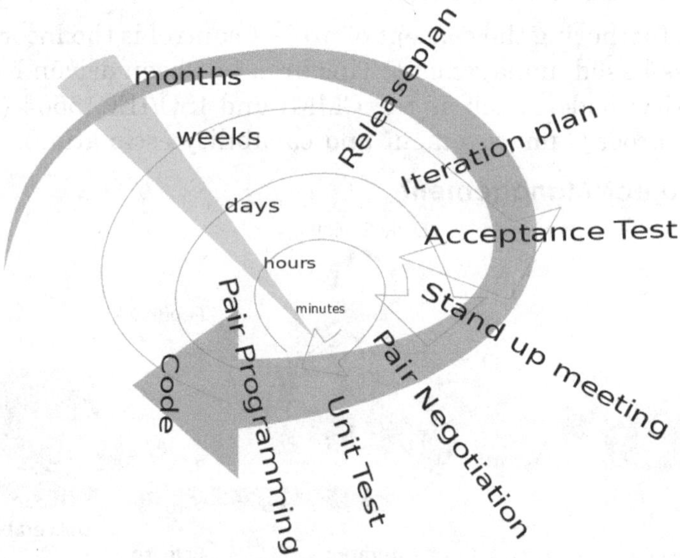

Figure: *Planning and feedback loops in Extreme programming (XP) with the time frames of the multiple loops.*

In critical studies of project management it has been noted that several PERT based models are not well suited for the multi-project company environment of today.

Most of them are aimed at very large-scale, one-time, non-routine projects, and currently all kinds of management are expressed in terms of projects.

Using complex models for "projects" (or rather "tasks") spanning a few weeks has been proven to cause unnecessary costs and low maneuverability in several cases.

The generalization of Extreme Programming to other kinds of projects is extreme project management, which may be used in combination with the process modelling and management principles of human interaction management.

Benefits Realisation Management

Benefits realization management (BRM) enhances normal project management techniques through a focus on outcomes (the benefits) of a project rather than products or outputs, and then measuring the degree to which that is happening to keep a project on track. This can help to reduce the risk of a completed project being a failure by delivering agreed upon requirements/outputs but failing to deliver the *benefits* of those requirements.

An example of delivering a project to requirements might be agreeing to deliver a computer system that will process staff data and manage payroll, holiday and staff personnel records. Under BRM the agreement might be to achieve a specified reduction in staff hours required to process and maintain staff data.

Processes

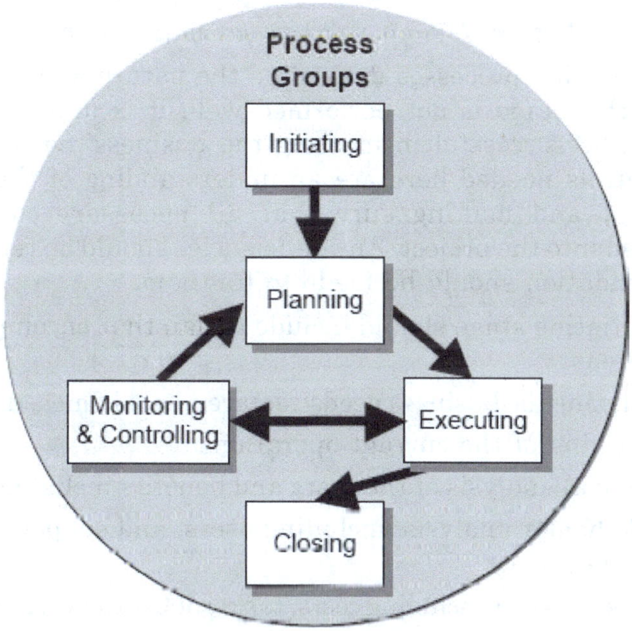

Figure: *The project development stages*

Traditionally, project management includes a number of elements: four to five process groups, and a control system. Regardless of the methodology or terminology used, the same basic project management processes will be used. Major process groups generally include:

- Initiation
- Planning or design

- Production or execution
- Monitoring and controlling
- Closing

In project environments with a significant exploratory element (e.g., research and development), these stages may be supplemented with decision points (go/no go decisions) at which the project's continuation is debated and decided. An example is the Phase–gate model.

Initiating

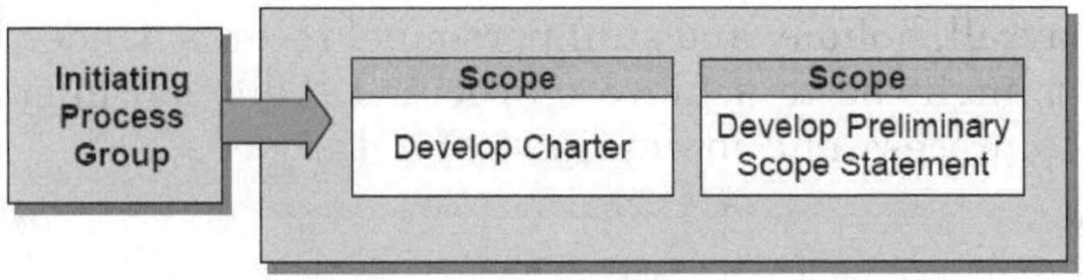

***Figure:** Initiating process group processes*

The initiating processes determine the nature and scope of the project. If this stage is not performed well, it is unlikely that the project will be successful in meeting the business' needs. The key project controls needed here are an understanding of the business environment and making sure that all necessary controls are incorporated into the project. Any deficiencies should be reported and a recommendation should be made to fix them.

The initiating stage should include a plan that encompasses the following areas:

- analyzing the business needs/requirements in measurable goals
- reviewing of the current operations
- financial analysis of the costs and benefits including a budget
- stakeholder analysis, including users, and support personnel for the project
- project charter including costs, tasks, deliverables, and schedule

Planning and Design

After the initiation stage, the project is planned to an appropriate level of detail. The main purpose is to plan time, cost and resources adequately to estimate the work needed and to effectively manage risk during project execution. As with the Initiation process group, a failure to adequately plan greatly reduces the project's chances of successfully accomplishing its goals.

Project planning generally consists of

- determining how to plan (e.g. by level of detail or rolling wave);
- developing the scope statement;
- selecting the planning team;
- identifying deliverables and creating the work breakdown structure;
- identifying the activities needed to complete those deliverables and networking the activities in their logical sequence;
- estimating the resource requirements for the activities;
- estimating time and cost for activities;
- developing the schedule;
- developing the budget;
- risk planning;
- gaining formal approval to begin work.

Additional processes, such as planning for communications and for scope management, identifying roles and responsibilities, determining what to purchase for the project and holding a kick-off meeting are also generally advisable.

For new product development projects, conceptual design of the operation of the final product may be performed concurrent with the project planning activities, and may help to inform the planning team when identifying deliverables and planning activities.

Executing

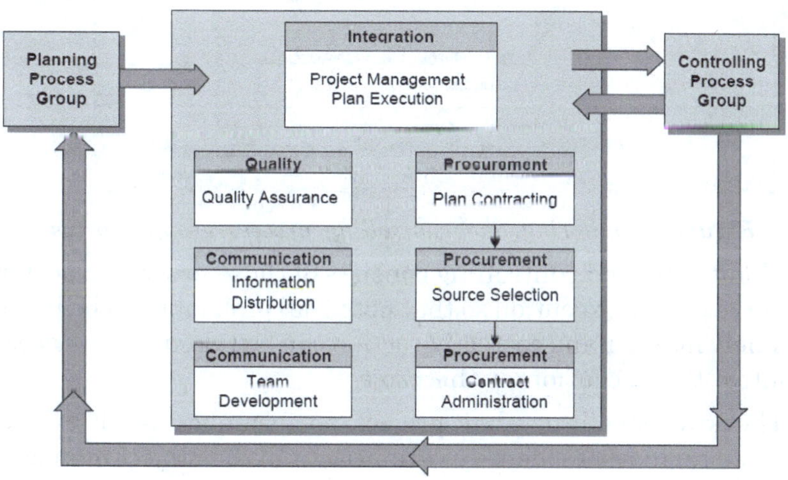

Figure: *Executing process group processes*

Executing consists of the processes used to complete the work defined in the project plan to accomplish the project's requirements. Execution process involves coordinating people and resources, as well as integrating and performing the activities of the project in accordance with the project management plan.

The deliverables are produced as outputs from the processes performed as defined in the project management plan and other frameworks that might be applicable to the type of project at hand.

Execution process group include:

- Direct and manage project execution
- Quality assurance of deliverables
- Acquire, develop and manage Project team
- Distribute information
- Manage stakeholder expectations
- Conduct procurement
- Test the deliverables against the initial design

Monitoring and Controlling

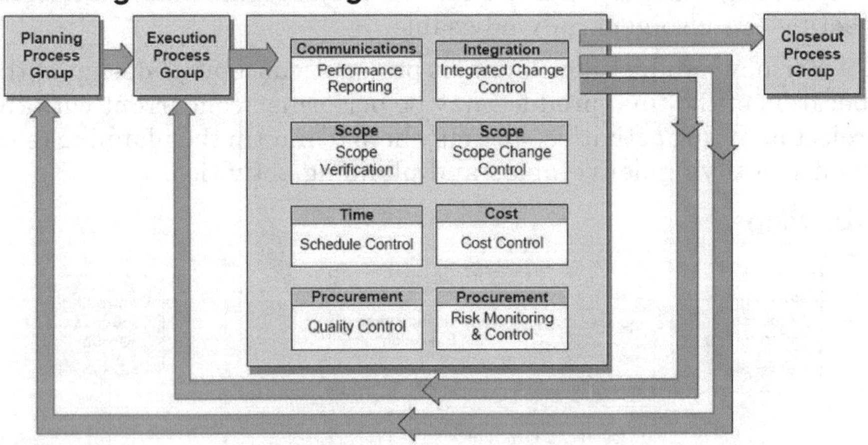

Figure: *Monitoring and controlling process group processes*

Monitoring and controlling consists of those processes performed to observe project execution so that potential problems can be identified in a timely manner and corrective action can be taken, when necessary, to control the execution of the project.

The key benefit is that project performance is observed and measured regularly to identify variances from the project management plan.

Monitoring and controlling includes:

- Measuring the ongoing project activities ('where we are');
- Monitoring the project variables (cost, effort, scope, etc.) against the project management plan and the project performance baseline (*where we should be*);
- Identify corrective actions to address issues and risks properly (*How can we get on track again*);
- Influencing the factors that could circumvent integrated change control so only approved changes are implemented.

In multi-phase projects, the monitoring and control process also provides feedback between project phases, in order to implement corrective or preventive actions to bring the project into compliance with the project management plan.

Project maintenance is an ongoing process, and it includes:

- Continuing support of end-users
- Correction of errors
- Updates of the software over time

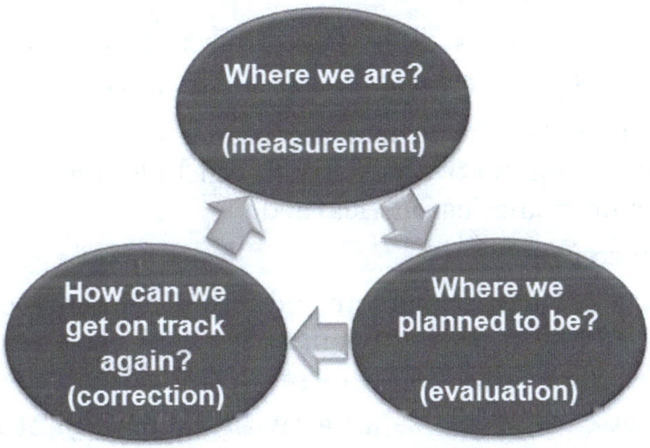

Figure: *Monitoring and controlling cycle*

In this stage, auditors should pay attention to how effectively and quickly user problems are resolved.

Over the course of any construction project, the work scope may change. Change is a normal and expected part of the construction process. Changes can be the result of necessary design modifications, differing site conditions, material availability, contractor-requested changes, value engineering and impacts from third parties, to name a few. Beyond executing the change in the field, the change normally needs to be documented to show what was actually constructed. This is referred to as change management. Hence, the owner usually requires a final record to show all changes or, more specifically, any change that modifies the tangible portions of the finished work. The record is made on the contract documents – usually, but not necessarily limited to, the design drawings. The end product of this effort is what the industry terms as-built drawings, or more simply, "as built." The requirement for providing them is a norm in construction contracts.

When changes are introduced to the project, the viability of the project has to be re-assessed. It is important not to lose sight of the initial goals and targets of the projects. When the changes accumulate, the forecasted result may not justify the original proposed investment in the project.

Closing

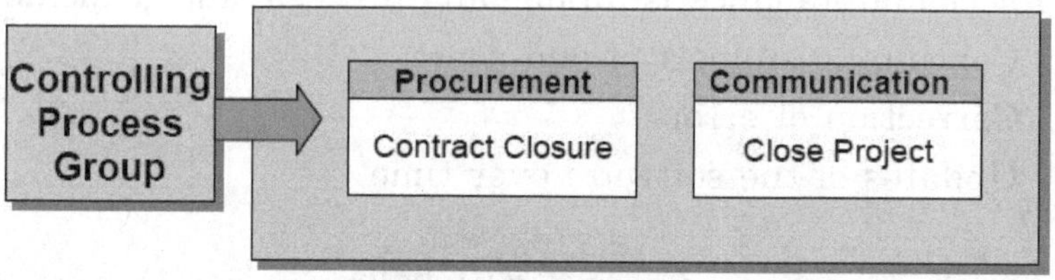

Figure: Closing process group processes.

Closing includes the formal acceptance of the project and the ending thereof. Administrative activities include the archiving of the files and documenting lessons learned.

This phase consists of:

- Contract closure: Complete and settle each contract (including the resolution of any open items) and close each contract applicable to the project or project phase.
- Project close: Finalize all activities across all of the process groups to formally close the project or a project phase

Project Controlling and Project Control Systems

Project controlling should be established as an independent function in project management. It implements verification and controlling function during the processing of a project in order to reinforce the defined performance and formal goals. The tasks of project controlling are also:

- the creation of infrastructure for the supply of the right information and its update

- the establishment of a way to communicate disparities of project parameters
- the development of project information technology based on an intranet or the determination of a project key performance index system (KPI)
- divergence analyses and generation of proposals for potential project regulations
- the establishment of methods to accomplish an appropriate project structure, project workflow organization, project control and governance
- creation of transparency among the project parameters

Fulfillment and implementation of these tasks can be achieved by applying specific methods and instruments of project controlling. The following methods of project controlling can be applied:

- investment analysis
- cost–benefit analyses
- value benefit Analysis
- expert surveys
- simulation calculations
- risk-profile analyses
- surcharge calculations
- milestone trend analysis
- cost trend analysis
- target/actual-comparison

Project control is that element of a project that keeps it on-track, on-time and within budget. Project control begins early in the project with planning and ends late in the project with post-implementation review, having a thorough involvement of each step in the process. Each project should be assessed for the appropriate level of control needed: too much control is too time consuming, too little control is very risky. If project control is not implemented correctly, the cost to the business should be clarified in terms of errors, fixes, and additional audit fees.

Control systems are needed for cost, risk, quality, communication, time, change, procurement, and human resources. In addition, auditors should consider how important the projects are to the financial statements, how reliant the stakeholders are on controls, and how many controls exist. Auditors should review the development process and procedures for how they are implemented. The process of development and the quality of the final product may also be assessed if needed or requested.

A business may want the auditing firm to be involved throughout the process to catch problems earlier on so that they can be fixed more easily. An auditor can serve as a controls consultant as part of the development team or as an independent auditor as part of an audit.

Businesses sometimes use formal systems development processes. These help assure that systems are developed successfully. A formal process is more effective in creating strong controls, and auditors should review this process to confirm that it is well designed and is followed in practice. A good formal systems development plan outlines:

- A strategy to align development with the organization's broader objectives
- Standards for new systems
- Project management policies for timing and budgeting
- Procedures describing the process
- Evaluation of quality of change

Topics

Project Managers: A project manager is a professional in the field of project management. Project managers can have the responsibility of the planning, execution, and closing of any project, typically relating to construction industry, engineering, architecture, computing, and telecommunications. Many other fields in production engineering and design engineering and heavy industrial have project managers.

A project manager is the person accountable for accomplishing the stated project objectives. Key project management responsibilities include creating clear and attainable project objectives, building the project requirements, and managing the triple constraint for projects, which is cost, time, and scope.

A project manager is often a client representative and has to determine and implement the exact needs of the client, based on knowledge of the firm they are representing. The ability to adapt to the various internal procedures of the contracting party, and to form close links with the nominated representatives, is essential in ensuring that the key issues of cost, time, quality and above all, client satisfaction, can be realized.

Project Management Triangle

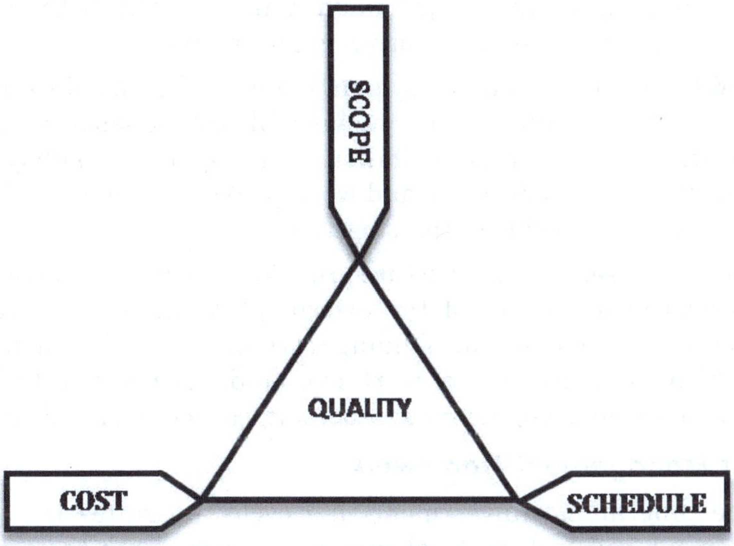

Figure: *The project management triangle*

Like any human undertaking, projects need to be performed and delivered under certain constraints. Traditionally, these constraints have been listed as "scope," "time," and "cost". These are also referred to as the "project management triangle", where each side represents a constraint. One side of the triangle cannot be changed without affecting the others. A further refinement of the constraints separates product "quality" or "performance" from scope, and turns quality into a fourth constraint.

The time constraint refers to the amount of time available to complete a project. The cost constraint refers to the budgeted amount available for the project. The scope constraint refers to what must be done to produce the project's end result. These three constraints are often competing constraints: increased scope typically means increased time and increased cost, a tight time constraint could mean increased costs and reduced scope, and a tight budget could mean increased time and reduced scope.

The discipline of project management is about providing the tools and techniques that enable the project team (not just the project manager) to organize their work to meet these constraints.

Work Breakdown Structure

The work breakdown structure (WBS) is a tree structure that shows a subdivision of effort required to achieve an objective—for example a programme, project, and contract. The WBS may be hardware-, product-, service-, or process-oriented.

A WBS can be developed by starting with the end objective and successively subdividing it into manageable components in terms of size, duration, and responsibility (e.g., systems, subsystems, components, tasks, sub-tasks, and work packages), which include all steps necessary to achieve the objective.

The work breakdown structure provides a common framework for the natural development of the overall planning and control of a contract and is the basis for dividing work into definable increments from which the statement of work can be developed and technical, schedule, cost, and labour hour reporting can be established.

Project Management Framework

The programme (investment) life cycle integrates the project management and system development life cycles with the activities directly associated with system deployment and operation. By design, system operation management and related activities occur after the project is complete and are not documented within this guide.

For example, in the US United States Department of Veterans Affairs (VA) the programme management life cycle is depicted and describe in the overall VA IT Project Management Framework to address the integration of OMB Exhibit 300 project (investment) management activities and the overall project budgeting process.

The VA IT Project Management Framework diagram illustrates Milestone 4 which occurs following the deployment of a system and the closing of the project.

The project closing phase activities at the VA continues through system deployment and into system operation for the purpose of illustrating and describing the system activities the VA considers part of the project. The figure illustrates the actions and associated artifacts of the VA IT Project and Programme Management process.

International Standards

There have been several attempts to develop project management standards, such as:

- Capability Maturity Model from the Software Engineering Institute.
- GAPPS, Global Alliance for Project Performance Standards – an open source standard describing COMPETENCIES for project and programme managers.
- A Guide to the Project Management Body of Knowledge from the Project Management Institute (PMI)
- HERMES method, Swiss general project management method, selected for use in Luxembourg and international organizations.
- The ISO standards ISO 9000, a family of standards for quality management systems, and the ISO 10006:2003, for Quality management systems and guidelines for quality management in projects.
- PRINCE2, PRojects IN Controlled Environments.
- Association for Project Management Body of Knowledge
- Team Software Process (TSP) from the Software Engineering Institute.
- Total Cost Management Framework, AACE International's Methodology for Integrated Portfolio, Programme and Project Management.
- V-Model, an original systems development method.
- The Logical framework approach, which is popular in international development organizations.
- IAPPM, The International Association of Project & Programme Management, guide to project auditing and rescuing troubled projects.

Project Portfolio Management

An increasing number of organizations are using, what is referred to as, project portfolio management (PPM) as a means of selecting the right projects and then using project management techniques as the means for delivering the outcomes in the form of benefits to the performing private or not-for-profit organization.

Project Management Software

Project management software has a capacity to help plan, organize, and manage resource pools and develop resource estimates. Depending the sophistication of the software, resource including estimation and planning, scheduling, cost control and budget management, resource allocation, collaboration software, communication, decision-making, quality management and documentation or administration systems. Today, numerous PC-based project management software packages exist, and they are finding their way into almost every type of business. Software may range from the high-end Microsoft Project to a simple spreadsheet in Microsoft Excel.

Virtual Project Management

Virtual programme management (VPM) is management of a project done by a virtual team, though it rarely may refer to a project implementing a virtual environment It is noted that managing a virtual project is fundamentaly different from managing traditional projects, combining concerns of telecommuting and global collaboration (culture, timezones, language).

Bibliography

Bilings, Henry: *Bridges*, The Viking Press, New York, 1956.

Bill, M.: *Robert Maillart: Bridges and Constructions*, Praeger, New York, 1969.

Brown, Allen: *Golden Gate: Biography of a Bridge*, Doubleday Press, Inc. New York, 1965.

Buckland, P.G.: *Esthetics in Concrete Bridge Design*, American Concrete Institute, Detroit, 1990.

Cassady, Stephen: *Spanning the Gate, The Golden Gate Bridge,* Square Books, Mill Valley, CA, 1986.

Figg, Eugene C.: *Proceedings of the International Bridge Conference*, Engineers Society of Western Pennsylvania, Pittsburgh, 2001.

Gauvreau, P.: *Innovation and Aesthetics in Bridge Engineering,* Canadian Civil Engineer, 2006.

Giedion, Siegfried: *Space, Time, and Architecture*, Harvard University Press, Cambridge, Mass., 1952.

Gottemoeller, Frederick: *Bridgescape: The Art of Designing Bridges,* John Wiley & Sons, NY, 2004.

Knapp, Edward John: *The Golden Gate Bridge Trivia Book*, Chronicle Books, San Francisco, CA, 1987.

Leonhardt, Fritz: *Brüken - Ästhetik und Gestaltung/Bridges - Aesthetics and Design,* MIT Press, NY, 1984.

Mock, Elizabeth B.: *The Architecture of Bridges*, The Museum of Modern Art, New York, 1949.

Owens, Tom: *The Golden Gate Bridge*, PowerKid Press, New York, 1997.

Robert Maillart: *Builder, Designer, and Artist. Cambridge*, Cambridge University Press, New York, 1997.

Salvadori, M.G.: *The Message of Structure, Engineering and the Humanities*, John Wiley & Sons, New York, 1982.

Schock, Jim: *The Bridge a Celebration: The Golden Gate Bridge,* Golden Gate International, Inc. Mill Valley, CA. 1997.

Sullivan, Louis: *Kindergarten Chats*, Wittenborn, Schultz, Inc, New York, 1947.

Torroja, E.: *The Beauty of Structures, Philosophy of Structures*, University of California Press, Berkeley, California, 1958.

Zuk, William: *Public Response to Bridge Colors*, Virginia Highway Research Council, Charlottesville, VA, 1973.

Index

❑❑❑